OUR

MINERAL

RESOURCES

NEW YORK · JOHN WILEY & SONS, INC.   London · Chapman & Hall, Limited

OUR
MINERAL
RESOURCES

CHARLES M. RILEY

AN ELEMENTARY TEXTBOOK IN ECONOMIC GEOLOGY

PREFACE

This book has been written with the viewpoint of supplying the needs for an elementary course in economic geology. Such a course could be an introduction to a curriculum for the training of a professional geologist, as well as a cultural course in earth science attended by students seeking degrees in other fields. It would normally follow freshman level courses in physical and historical geology, although the latter is not necessary for an understanding of the subject matter treated in this book. A knowledge of high-school or college chemistry is very desirable, but I realize that many students at this level would not have such a background. Where discussions of geologic phenomena require an explanation of chemical processes, the necessary principles of chemistry are presented in an elementary manner. However, the student should be well informed on the subject matter of his course in physical geology, including such topics as the nature of common rocks and minerals, the formation of folded and faulted rock structures, volcanism, and the effects of weather on earth materials.

The book is divided into two sections, the first concerned with the geologic occurrences of the common metals used by man, and the second with the geology of nonmetallic resources. With the exception of Chapter 1, each chapter is devoted to a metallic or nonmetallic resource in common use today.

v

Into Chapter 1 I have condensed nearly all the geologic principles which govern the formation of a majority of metallic ore deposits and many of the nonmetallic deposits. Most metallic occurrences fall into a widely accepted genetic classification which in a condensed form in Chapter 1 has been used as an outline for descriptions of the many types of deposits and of the earth processes responsible for their formation. The mineral deposits which serve as examples of the natural occurrence of the metals and nonmetals discussed in later chapters are classified in the same way. Chapter 1 also presents a brief review of the history of ideas which scientists have proposed for the origin of mineral occurrences. It was felt that only by such an introduction could the reader appreciate the present-day controversy about the formation of many primary lode deposits. Some of the newer theories are summarized, but the limited space and the elementary level of this book prohibit a thorough debate of these ideas. It is hoped that this chapter, even with its abbreviated treatment of a difficult and controversial subject, will provide the necessary concepts for an understanding of the origin of mineral deposits.

New terminology and technical words plague every beginning student in geology. It is impossible to discuss a scientific subject without using some technical terms, but their use in this book has been kept to a minimum. A glossary has been appended to explain the terms which are not in the average student's vocabulary but which are necessary for a simple explanation of the subject matter. The names of geologic eras, periods, and epochs and their time duration; the chemical elements and their symbols; and common minerals and their compositions have also been added as appendices.

I wish to extend thanks to many people who contributed to the successful completion of this book. Professor A. L. Lugn of the University of Nebraska reviewed the chapter on ground water and made many helpful suggestions. The chapter on aluminum was read by the technical staff of the Aluminum Company of America; their helpful comments added much to its clarity and accuracy. Professors R. L. Threet of the University of Utah and J. H. Fisher of Michigan State University willingly gave of their time to assist me on many occasions. I am grateful to the University of Nebraska and the Department of Geology for the encouragement which they have given as well as material aid they provided during the years of preparation of this book. No little thanks are due Mrs. Joan Braden, secretary in the Department of Geology, who carefully typed the entire manuscript. Many companies kindly donated photographs which add much to the value of the book. I am indebted to the concerns whose names appear with the photographs they provided. Most of all, I am grateful to my wife, Ellen Nadine Riley, without whose constant encouragement and inspiration this book would never have been completed.

It is impossible to acknowledge every source of information used in the writing of my book. I relied heavily on the books listed as General References as well as the many selected references at the end of each section. The authors of these works have not been cited specifically in the text for reasons of simplicity, but I am nonetheless grateful for their contributions to the science as used in this book.

CHARLES M. RILEY

May, 1959

CONTENTS

SECTION II   NONMETALLIC MINERALS

introduction

The average citizen gives little thought to the many cultural and technological advances that play so great a part in the high standard of living which he is enjoying. Evolution of language, religion, ethnology, and philosophy through the ages has created a social environment for Mr. Citizen which he feels is essentially right. Medical advances have given him a life expectancy which is double what the average worker could expect in the Middle Ages. Great strides in the knowledge of agriculture, animal husbandry, and food preparation have given our modern man a diet of a quality and variety never before enjoyed on this earth, even by the richest kings of old. If we were to ask Mr. Citizen about his standard of living, he might mention some of these less-obvious blessings, but more likely he would talk of the wealth of physical and mechanical things which make his life easy and enjoyable. His modern home, automobile, refrigerator, radio, television, telephone, electric lights, and a host of other modern inventions would probably be first on his lips. It would, perhaps, not occur to him to think further to the raw materials without which he could have none of these necessities and luxuries of his modern world. Let us point out some of the basic mineral substances which he uses daily without realizing it.

In the construction of his home a variety of metals are used, including iron, copper, zinc, lead, aluminum, tungsten, and, perhaps, alloys containing manganese, chromium, nickel, and molybdenum. Stone or bricks may be used on the outer surface, or perhaps a shingle siding made of asbestos and cement. The asphalt roofing paper is covered with stone chips, and the inner walls are of plasterboard, made mostly from gypsum. Glass is generously utilized in his modern home, but it must be manufactured from quartz, soda, lime, and a variety of other materials. The basement and footings are concrete in which is cement (limestone, clay, gypsum) mixed with sand and gravel. The house is insulated with mineral glass fiber and is painted a snowy white with paint that contains lithopone (zinc sulfide and barium sulfate mixture) and titanium oxide pigment. His heating plant burns natural gas, but his friends use fuel oil and coal, all mineral fuels. Water is piped to his house from a municipal system that relies upon surface water or ground water sources. His electricity is generated in a plant which burns coal or gas, and it travels to the many outlets in his house through networks of copper wire.

When one starts to enumerate for Mr. Citizen all of the mineral raw materials used in his automobile or the train in which he rides to work, the list becomes even more impressive. In his office building and in the city around him are things made with mineral products from mines, wells, and quarries all over the world. Even his food and clothing, products of agriculture, would not be as good or as inexpensive were it not for the great quantities of mineral fertilizers used each year by the farmer. In the following pages the reader will be constantly reminded of the great quantity and variety of minerals that are required in our complex industrialized world.

DISTRIBUTION OF MINERAL RESOURCES

Any student of economic geography is soon impressed with the fact that for every important mineral resource there are "have" and "have-not" nations. It is necessary that the nations with an abundance of a certain mineral share this resource with less-fortunate countries through the avenues of international trade. This is not done through any altruistic motive, for in return the supplier nation receives minerals or goods which it cannot produce within its political boundaries. There is such a variety of mineral raw materials needed today and they are so unequally distributed on this earth that no country is completely self-sufficient. Even the great industrial nations, United States, England, and Russia must rely upon imports of many minerals from all over the world.

United States has become a world power largely because of her bountiful mineral resources, which include nearly 50% of the world's reserves of coal, about 30% of all the iron ore, 25% of the copper, 50% of the phosphate rock, and abundant reserves of petroleum, sulfur, zinc, lead, and other vital minerals. The United States is producing most of these substances at a prodigious rate for domestic consumption and for export, yet she must import additional supplies of some others to meet the growing demands of her industries. There are many essential mineral substances for which she is wholly or almost entirely dependant upon imports. Tin, nickel, manganese, chromium, industrial diamonds, quartz crystals, and sheet mica are a few of these.

The mineral resources of the U.S.S.R. may ultimately prove to be more varied and plentiful than those in America. There are great areas that have been only poorly examined, and many regions are so remote that mineral resources they contain could not easily be exploited at the present time. However, the potentialities are great, and, even now, known deposits of iron ore, manganese, chromium, coal, mineral fertilizers, and a variety of other metallic and nonmetallic minerals are sufficient for her present needs and for export. The possibilities for extensive petroleum reserves are good, but little has been found so far. U.S.S.R. is deficient in and must import such mineral commodities as copper, lead, zinc, mercury, sulfur, tin, and many others.

In contrast to the two great nations best endowed with mineral wealth, there are the many countries which may have a few or several mineral substances in great abundance but do not have the industrial development to make use of them. Many backward but mineral-rich African and Asian countries are in this position, suppliers of raw materials and markets for manufactured goods.

The unequal distribution of minerals on the earth has in the history of mankind led to wars of conquest, colonialization, and purchases of vast undeveloped holdings by other countries. With mineral wealth has come economic power, particularly since the Industrial Revolution when coal and iron ore became the major factors in the industrial growth of a nation.

The days of colonialization are past, and no longer will nations sell lands to other countries. Unfortunately, wars of conquest still take place, but now the issues involved are more complex than the simple desire to steal the resources of a neighbor.

MILITARY IMPORTANCE OF MINERAL RESOURCES

War is a reality that seems to be threatening or with us all the time, despite a seemingly universal desire for peace. In the last 20 years the world has seen at least four wars as well as several revolutions and "police actions."

In olden times a country could train and simply equip a few thousand soldiers who could live mostly off the land. Such an army could win a war with a neighboring country by its fighting skill and the clever leadership of an experienced general. Modern wars can only be won by countries who produce the most and the best armament and military material of all kinds. Valorous soldiers and excellent leadership are important, but great industrial potential is essential, coupled with the maximum efforts of soldiers and civilians alike. Such military efforts create extraordinary demands for minerals. Metals of all kinds are consumed in vast amounts during wartime, and the drain on the world's reserves of mineral fuels is increased many times in order to keep the factories producing at maximum capacity and to propel the many military ships, planes, and vehicles. In short, a country with the most military power is one with the greatest available industrial potential, and this is greatly dependent upon the mineral resources which it controls. History has showed that a large well-trained and even well-equipped military machine is not enough to win a major war against an enemy with higher industrial potential. Germany, Japan, and Italy learned this again in World War II.

In recent years it has become more and more apparent that a new factor must enter into a determination of the military strength of a nation. This is the number and quality of trained scientists which the nation can direct to studies and research on new weapons and defenses. It was a cooperative scientific effort on a large scale that created the first atomic bomb, and today teams of scientists in many nations are perfecting the terrible hydrogen bombs, working on guided rocket missiles powerful enough to penetrate outer space, and developing other new weapons that we will perhaps never hear of until they are used.

Of course, industrial or military developments of this kind result in shortages and demands for new mineral materials. The discovery of atomic weapons created an unheard of demand for uranium, a metal that was previously considered little more than a curiosity yet was a rather plentiful by-product from the recovery of radium. Extensive explorations for uranium deposits by trained geologists and hopeful prospectors reached into every corner of the earth. The search was stimulated by government subsidies and other enticements. Many new discoveries were made, and this potent metal is flowing into the atomic arsenals of the world in greater and greater tonnages. Now an evaluation of military might must take into account the tonnage of uranium ore owned or controlled by a nation and the ability of the nation to transform the raw material into nuclear weapons.

EXHAUSTIBILITY OF MINERAL RESOURCES

Our agricultural resources can be considered everlasting, provided that we take care of the soil. Fields will yield crops year after year, and even forests can be cut again within a few decades. This is not true of our mineral resources. When a copper deposit is mined out, all that remains is a hole in the ground, and there is no possibility that more copper ore will form in the same place. Many great ore deposits have reached this state, and ghost towns are all that are left of what were once bustling mining communities. As long as new deposits are found to replace those that are exhausted, industry will continue to get the minerals it needs. For most minerals geologists and prospectors have been able to replenish our proven ore reserves by new discoveries. However, the rate of consumption of many of these minerals has increased so much in the last 20 years that our reserves will completely vanish in a short time if we fail to uncover even more and larger deposits. None of us is so optimistic as to believe that we can continue finding minerals as fast as we consume them. After the more exposed deposits are used, the deeper deposits can be discovered only by much financial risk and expensive exploratory programs. Not even the greatest of human ingenuity can find new ore deposits where none exist.

No one can exactly foretell when supplies of minerals will run out, because we do not know how many new discoveries will be made or what the demands of the future will be. Many predictions have been made in recent years, all taking into account the possible relief of the situation by new discoveries and by diminished demands upon mineral reserves when they become critically low.

Although these experts do not agree in their forecasts, they are all in accord that the long-range outlook is gloomy and that within a few centuries we are going to be completely without, or suffering severe shortages of, some minerals which are now considered essential to our economy.

The mineral fuels and many nonmetallic minerals present the most serious problems because they are "expendable" and cannot be used more than once. Coal is the mineral fuel that occurs in the greatest abundance, but even it will be depleted in about 2000 years if used at the present rate. Long before this, however, the costs of mining deeper less-profitable deposits may make it impossible for coal to compete with other sources of energy. The position of the world with regard to petroleum is more serious, because oil products are being consumed at an ever-increasing rate. In 1946 the world was burning *each day* 7,778,000 barrels of petroleum. In 1955 the consumption was 16,000,000 barrels per day, and in 1957 the rate was 17,500,000 barrels. New discoveries continue to keep up with the rate of depletion of known reserves, but exploration costs become greater as less-promising areas are tested. Even at today's consumption rate 200 years may see us without oil.

Nearly all minerals are required in greater quantities today than ever before. This is due to increased industrialization, higher living standards, and larger population. The United States alone, since 1914, has used more metals and mineral fuels than were used by all of the world in historic time preceding 1914.

There is no solution to the problem of the world's exhaustible mineral resources. We can delay the day of reckoning in many ways, but, so long as we use any mineral material, a time will come when there will be none left or when it will become uneconomical to recover it.

There are four things that can be done to stave off the unpleasant realities of this problem: (1) We must increase our programs of exploration, train more and better geologists to carry them out, and employ and develop the best geophysical and geochemical tools to aid in the search. (2) For the nonexpendable minerals we can increase the efficiency of scrap recovery so that these materials may be reused to a greater extent with less waste. (3) Researchers must find substitutes for each mineral when it can no longer be produced in great enough quantities or at low enough costs to meet the demands of industry. (4) We can restrict the uses of some minerals to those applications where they are most critical.

Substitution seems to be the only way man can forestall or significantly postpone the day when an industrial mineral will no longer be available for human use. Already many important advances have been made in a quest for other sources of energy to take the place of our mineral fuels. Solar energy has been utilized on an experimental basis, and it is entirely conceivable that much of our electrical power will some day be generated by great plants located in the arid regions of the world where there is a high incidence of sunshine. Solar energy may even be used to distill fresh water from sea water which will help create productive agricultural regions out of what now are deserts. Studies are also being made to find ways of harnessing some of the enormous energy of the tides, and increased usage of river water power will certainly take place.

Uranium is a new source of energy that is already being used as a substitute for coal and oil. Power from the atom will become more important in the next few decades, yet, as a substitute, it can only prolong our use of mineral fuels, for it too is an expendable mineral resource. Of course, any substitute for coal and petroleum should be welcome, for these fuels of today may mean far more to mankind as sources of organic chemicals than as sources of energy.

The most important kinds of substitute materials for our diminishing mineral resources are those that come from agriculture and are thus replenishable or those that are minerals so abundant in nature that we need never fear for their depletion. Plastics offer examples of agricultural substitutes that have taken the places of metals in a number of uses. Metals, ceramic materials, and rock products produced from sand, gravel, clay, limestone, and other plentiful rock materials must be used wherever possible, for such sources of mineral substitutes can never be used up. Minor amounts of a number of valuable metals and nonmetals occur in common granite, a source that could supply our needs indefinitely if only there were ways to extract these minerals economically. The ocean is an inexhaustable source for a variety of mineral substances such as magnesium that will become important substitutes in the future.

SELECTED REFERENCES

Ayres, Eugene. The fuel situation, *Scientific American*, Vol. 195, No. 4, pp. 43–49, 1956.

Leighton, M. M. *Our natural resources: their continuing discovery and human progress,* Ohio Div. Geol. Surv. Inf. Circ. No. 12, 16 pp., 1953.

Nolan, Thomas B. The outlook for the future—nonrenewable resources, *Econ. Geol.,* Vol. 50, No. 1, pp. 1–8, 1955.

Paley, William S., *et al. Resources for Freedom,* The Report of the President's Materials Policy Commission, 5 volumes, U.S. Govt. Printing Office, 1952.

Van Royen, W., and Oliver Bowles. Atlas of the World's Resources, *Mineral Resources of the World,* Vol. II, University of Maryland, Prentice-Hall, Englewood Cliffs, 1952.

metallic

section one

mineral resources

principles
of
ore deposition

CONCENTRATION

The rocks and soil beneath our feet, the air, the oceans, and all living things are made of only about 90 chemical elements. All matter is composed of these same fundamental materials which may be combined in mineral and organic compounds or occur free as gases, nonmetallic substances, and native metals. The relative abundance of the elements in the crust of the earth was computed in 1924 by Clarke and Washington, two famous geochemists of the U. S. Geological Survey. Since no one yet has any direct knowledge of the chemical nature of the interior of the earth, these men considered only the outer shell 10 miles thick. The figures were computed from thousands of chemical analyses of rocks sampled from all over the world. It was estimated that 95% of this outer shell of the earth is igneous rock, 4% shale, 0.75% sandstone, and 0.25% limestone. The atmosphere and oceans comprise less than

0.037% of the weight.   Their conclusions, reproduced in part below, were strikingly verified by subsequent investigations.

PERCENTAGE BY WEIGHT OF ELEMENTS IN THE EARTH'S CRUST

oxygen	46.60%	sodium	2.83%
silicon	27.72	potassium	2.59
aluminum	8.12	magnesium	2.09
iron	5.00	titanium	0.44
calcium	3.63	hydrogen	0.14
		all others	0.83

Some startling observations can be made from these figures.   Only 10 elements make up over 99% of the earth's crust, and oxygen and silicon alone comprise nearly 75%.   As silicates[1] are the most abundant minerals in all rock types, this is not too surprising.   Iron and aluminum are the only elements of the first ten that have been extensively used as uncombined metals.   Most of the useful metals and nonmetals with which we will be concerned in the following pages are part of that less than 1%.   It is also noteworthy that the important nonmetallic elements, carbon, chlorine, phosphorus, nitrogen, hydrogen, sulfur, and many familiar metals, occur in such sparse quantities.   The explanation for their ample supply lies, of course, in their great local abundance in certain parts of the earth's crust.

Concentrations many times higher than their average crustal amounts are required before most elements can be utilized.   Consider copper for an example: If all the elements were equally distributed, the average universal rock would contain approximately 0.007% copper.   It would be necessary to send a little over 7 tons of rock to the smelter to recover only 1 pound of copper.   Other common metals are even more scarce (that is, lead, 0.0016%; tin, 0.004%; uranium, 0.0004%; silver, 0.0001%; and gold 0.000005%).   It is obvious, therefore, that usable occurrences of these metals in the crust of the earth must have been created by natural processes of concentration.

ORE

A natural mineral occurrence that is sufficiently concentrated to be profitably utilized as a source for one or more metals is called an

---

[1] Silicates are chemical compounds of silicon, oxygen, and some metal.

ore. Usable deposits of some nonmetallic substances such as gems and fluorspar are also called ores. Any natural occurrence might be a "mineral deposit," but only one that can be exploited at a profit is an ore. This is, therefore, both a geologic term and one in the realm of economics. The profit from a mining operation is affected not only by the concentration of the valuable mineral in the ore but also by the efficiency of the operator; by the costs of labor, transportation, and equipment for mining and milling; by available markets; and finally by the value of the product. A deposit favorably located with respect to markets, transportation facilities, and labor sources may be an ore; whereas the same deposit located in a remote area would not be. A sharp drop in the market value of a metal may leave a marginal operation without material that can be worked profitably—a mine with no ore. On the other hand, a rise in the market could double its ore reserves by allowing the utilization of lower grade material previously considered submarginal.

ORIGIN OF ORE DEPOSITS

The economic factors that are involved in the definition and evaluation of ore bodies are of great concern to the mining geologist when he is called upon to evaluate mineral properties. He is also met with the difficult problems of determining the geologic nature of the mineral occurrences and of reconstructing the earth processes that led to their formation. The origin of each new ore deposit is an important problem that, when solved by the geologist, better enables him to guide the mining of the ore and even helps him to locate other deposits of a similar nature. For most ore deposits the geologist can determine with reasonable certainty the nature and time of their origin, but some are so complex or the evidence so incomplete that he can make only speculative deductions about their origin. Controversies about such deposits have been many and heated, but most have been resolved by the discovery of new evidence or the recognition of new concepts.

Science progresses by the clash of divergent ideas, for the prolonged debates stimulate further research and ultimately lead us a little closer to the truth.

Certain natural processes are active today at the surface of the earth and can be easily observed. Mineral deposits resulting from such processes as sedimentation and weathering afford no great mysteries for the geologist when he can watch them form. Those processes such as vein formation and igneous intrusion which occur deep beneath the surface can only be inferred from the evidence left in the rocks. It is deposits ascribed to such activities over which most of the debate has centered. We can often deduce the temperature and depth of formation of such occurrences, but difficulties arise in the determination of the source of the metals, their mode of migration, and the reasons for their accumulation.

Problems such as these are today being attacked in the laboratories of the geochemists who are making great strides in determining the physical and chemical environments in which various key minerals can form. The answers will surely be found by combining the newly discovered geochemical concepts with careful geologic field studies.

History of Theories on Ore Formation

The many theories on the formation of ore deposits which have evolved since man first engaged in mining are far too numerous to mention here. The major advances have occurred mostly in the last century, but there are few "new" theories that have not been expressed earlier in a rudimentary form. Men such as Theophrastus (372–287 B.C.), Avicenna (A.D. 980–1037), Agricola (1494–1555), Descartes (in 1644), Werner (1749–1807), Hutton (in 1788), and de Beaumont (in 1847), proposed ideas that were the foundation stones upon which our science was built.

Near the end of the nineteenth century a theory of ore deposition was proposed that was known as *lateral secretion,* which means deposition of lodes from water that percolated horizontally a long way through the rocks. The water was originally rain that soaked downward, became heated, and dissolved minor amounts of metals from the rocks as it traveled.

Although lateral secretion was advocated by many respected geologists, it was strongly contested by others who held that ascending magmatic *hydrothermal* (*hydro* water + *thermal* heat; that is, "hot water") solutions deposited the vein ores. This concept, known as the *ascentionist theory,* won out in the heated debate that finally subsided by the beginning of the twentieth century. For a few decades there was general accord in the belief that magmas and ascending hydrothermal fluids derived from magmas were responsible for most ore deposits.

There have always been weak points in the magmatic-hydrothermal theory of ore deposition, and many ore deposits are known for which it does not fully explain the facts. Consequently, in recent years new hypotheses have arisen that have attracted many advocates among our leading scientists. One of these might be called the *neolateral secretion* theory, for it resembles strongly the ideas debated at the end of the last century. It proposes that rare metals, present in nearly all rocks in minor amounts, may be mobilized under conditions of intense metamorphism and redeposited in suitable host rocks and structures as ore bodies. The metals are said to move by diffusion through water-filled rock pores or even to migrate as ions without the aid of liquids. Water in pores or chemically combined in the source rock is driven out by the metamorphism. It moves as a wave ahead of the intense heat, forms veins, and causes many of the rock alterations around veins. The ultrametamorphism, of which ore formation is just one aspect, is a process capable of adding and removing much chemical material, and by recrystallization it is said to be able to change large masses of sedimentary rocks into the great granitic bodies known as batholiths. This process is known as *granitization*. Proponents admit that magmas do occur but attribute their origin to melting caused by overintense granitization.

It has long been recognized that certain metals can be deposited with sedimentary rocks in quantities sufficient for mining. Other metals are known to occur in sediments in minor or trace amounts. It has been proposed that the metamorphism or granitization of such rocks would concentrate the metallic elements. The resulting ore bodies must then show a spatial and genetic relationship to one another and to the parent beds. This application of neolateral secretion to rocks already rich in metals has been used to explain mining districts of great extent where many similar deposits contain the same metals.

CLASSIFICATION AND EXPLANATION OF METALLIC ORE DEPOSITS

By now the reader should be well aware that geologists are not yet in agreement on some very fundamental aspects of ore formation. No theory is adequate to explain all deposits, and it is beyond the scope of this book to debate the pros and cons of even the leading theories. Therefore, the still widely accepted magmatic-hydrothermal theory is used as a basis for the following classification and for the explanations of origin of many types of deposits.   Where appropriate in various sections of the book and particularly in this chapter, conflicting views are summarized.

Classification of Ore Deposits

Of the many classifications devised for ore deposits, the ones based upon modes of origin have been most useful.   Such classifications require that the geologist study all the facts available for the deposit under consideration and interpret these facts from the wealth of geologic knowledge, geologic theory, and his own experience.   Difficulties in classification arise when sufficient facts are not available or through the inherent failure of any classification of nature to provide enough categories or to allow for gradations.   The following classification is admittedly oversimplified, but for the beginner in the study of economic geology, it provides enough major groupings to include most of the important ore deposits.   The remainder of the chapter describes in order the characteristics of these different genetic classes.

Genetic Classification of Ore Deposits

1. Magmatic deposits
   a. Disseminated magmatic deposits
   b. Deposits formed by early crystal settling
   c. Deposits formed by filter-press concentration
   d. Late magmatic deposits
2. Pegmatites
3. Contact metamorphic deposits
4. Hydrothermal deposits (veins, replacement lodes, etc.)
5. Sedimentary deposits
   a. Chemical precipitates
   b. Placer deposits
6. Deposits of secondary enrichment
   a. Residual concentrations
   b. Supergene enrichment

The Role of Magma in Ore Formation

What is a magma?   Most definitions describe magma as a very hot liquid in the earth which, upon cooling, forms glassy or crystalline igneous rocks.   When magma flows onto the surface, it becomes *lava*. A greater insight into the nature of this "very hot liquid" is necessary before one can appreciate the important role which magma has seemingly played in the formation of most ore deposits. Because no one has ever seen a magma, we can only deduce its characteristics from the igneous rocks and other features that are said to result from magmas. Even less is known about its origin, and the many theories belong in the realm of philosophical speculation.

Suppose that a piece of granite were heated in a high-temperature furnace until it completely melted. Would this molten material be a magma? The answer is no, and for this reason: When the original piece of granite formed from a magma, the biotite mica, orthoclase feldspar, and quartz crystallized in turn to form the granite. However, many constituents of that magma remained liquid during the crystallization of the granite minerals and did not become a part of the igneous rock. These residual constituents were very fluid and would have quickly volatilized at the high temperatures had it not been for the pressure. Because the granite formed perhaps a mile or more below the surface of the earth, it was under high pressure from the weight of the overlying rock. This great pressure caused the volatile constituents to migrate through available rock openings toward the surface where the pressure is low. For this reason, some of the original magma did not become part of the granite but was separated from it during the cooling process. As all magmas are believed to contain these volatile substances, the melting of a piece of granite, therefore, yields only the nonvolatile part of the magma from which the granite crystallized. Even by slow cooling the experimental melt will not solidify as granite in its original state.

What are the volatile constituents? We must identify the volatile constituents of a magma and understand the role they play in its crystallization. We will also see how the volatiles cause the formation of ore deposits that are closely associated with the igneous rock.

Although the volatile substances in magmas are seldom represented in the chemical constituents of the igneous rocks, it is still possible to determine what some of them are. Where magmas come to the surface of the earth in the craters of volcanoes, the sudden decrease in pressure allows the gases to escape. This is illustrated by a bottle of soda pop that has carbon dioxide gas in solution. While the cap is on the bottle, the pop is under pressure and no gas can escape; but remove the cap and the carbon dioxide quickly appears as tiny bubbles which rise to the surface. After a while the pop has lost most of its dissolved gases and is "flat." The magma entering the crater of a volcano loses its gas for the same reasons. By collecting the escaping volcanic gases and analyzing them, the scientists can tell directly what the volatile constituents of that magma are. Studies of volcanoes in Italy, Hawaii, the Aleutians, and other regions of active volcanism have shown that over 90% of the gas given off by magmas is water vapor. Also present are oxygen, nitrogen, carbon dioxide, hydrochloric acid, hydrofluoric acid, boron, and various gases of sulfur. It is interesting to note that small amounts of many metals including iron, copper, lead, tin, zinc, nickel, cobalt, barium, manganese, and even gold are carried by these gases. This is compelling evidence in support of the theory that many metallic ore deposits are formed in some manner through the agency of the volatile constituents of magmas.

Those who believe in granitization propose that volatile material similar to that described above is driven out from the sediments during intense metamorphism. Following these volatiles, and in part dissolved in them, may be the great quantities of chemical substances that must be removed from sedimentary rocks before granitization is completed. Waves of these fluids, rich in calcium and magnesium, are said to alter the rocks ahead of the advancing granitization and has been alluded to as "the basic front." The advance waves of heated water solutions of the basic front are the ore-forming hydrothermal solutions of granitization. Intense granitization is said to cause melting with the production of magma that may crystallize as small truly igneous bodies or become the source for volcanic flows of lava.

No one can contest in the light of recent discoveries that granitization does exist, but the divergence in ideas arises in whether batholiths can be created by the process without the formation of an original magma. As we do not know what is the source for magma, perhaps the granitizationists and the magmatists are not at such opposite poles but are arguing about different parts of one continuous process. One group is impressed with the processes that may lead to the creation of magma, and the other is concerned with the events that take place during the cooling of the magma. Perhaps, ore deposits can occur during both stages, but one cannot escape the evidence that a preponderance of ore deposits is closely and genetically associated with clearly magmatic igneous bodies.

MAGMATIC ORE DEPOSITS

Because of the strong indications that most ore deposits are directly or indirectly related to magma, the classification "magmatic ore deposit" is confusing. However, general usage has given it the following meaning: *a deposit that has originated from a magma and is an integral part of the igneous rock formed from that magma.* This definition excludes those ore deposits that result from the volatile constituents of the magma.

Magmatic deposits form by several different processes which are thought to take place during the cooling and crystallization of a magma. A few of the more important types of ore deposits formed by these processes can be classified in the following way:

Disseminated magmatic deposits. In these deposits the ore minerals have crystallized as part of the igneous rock and are disseminated through it, usually as isolated crystalline grains. No natural processes have concentrated these minerals. Many useful metallic minerals occur in this way but make up so low a percentage of the rock that they could never be recovered at a profit. For this reason, such deposits could not be considered ores. Diamonds, however, occur as disseminated crystals in certain kinds of igneous rocks. Because of the great value of this mineral, the whole igneous rock mass can be mined and the diamonds recovered at a profit. In this case all of the igneous rock is ore.

Early crystal settling. When a magma cools, certain minerals crystallize from the liquid at a higher temperature than the common igneous rock minerals. If these early crystals are of a substance that is denser than the magma, they will sink to the bottom of the magma chamber. The accumulation of such crystals results in a concentration of one or more of the chemical substances which existed in very small proportions in the parent magma. If such substances are useful to man, a valuable ore deposit may be created by this process.

"Filter-press" concentrations. Picture a magma slowly cooling deep within the earth (Fig. 1). Some early high-temperature crystals have already formed. These crystals either are suspended in the remaining liquid or are settling slowly toward the bottom of the chamber. Now compressional forces in the earth may squeeze and contract the magma chamber with the result that the remaining liquid is forced out of the chamber through small cracks or fissures. If the crystals are too large to pass through these openings, a very effective separation results between the early formed crystals and the remaining liquid. This process might be compared with the action of an old-fashioned wine press. The grapes are placed in a large wooden container with small holes or cracks in it. Pressure is applied by a piston from above, and the juice is forced out the small openings, leaving the pulp behind. In the case of the mag-

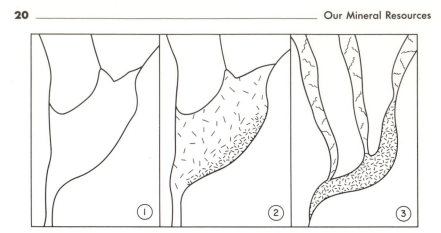

Fig. 1.   Progressive stages in the formation of a magmatic ore deposit by filter pressing: (1) first injection of magma, (2) early crystallization of ore minerals, (3) remaining liquid squeezed out into dikes leaving ore concentration.

ma, if the early formed crystals are of some valuable substance, this separation or concentration creates an ore deposit in the original magma chamber.   If most of the igneous rock minerals have formed in the magma chamber and some valuable substances are concentrated in the remaining liquid, the filter pressing will cause valuable ore deposits to be formed by the residual magma in the cracks and fissures surrounding the original magma chamber.

Late magmatic deposits.   More detailed studies of magmatic ore deposits, some long considered to be early magmatic, have shown that in the majority of cases, the ore minerals have crystallized late in the cooling history of the magmas.   The ore solutions are believed to be highly fluid residual magma, perhaps concentrated by filter pressing or some other mechanism.   Some of the volatile constituents of the magma may play a part in the formation of these deposits.

Other mechanisms have been suggested for the segregation of valuable substances from magma, either from the early crystallizing constituents or from the late residual "juices," but the resulting mineral deposits can be recognized as magmatic ores by the following criteria:

1. They may occur as streaks, layers, or pockets near the bottom of an intrusion where they are completely surrounded by and grade into the igneous rock formed from the same magma.

2. They usually contain an appreciable content of minerals common to the parent igneous rock.

3. If not an actual part of the parent igneous rock, they should always be very near to it and connected to it.

4. They display structural and textural features common to coarse-grained igneous rocks and never have the characteristics of fine-grained extrusive igneous rocks.

5. They are most commonly associated with the dark ultrabasic igneous rocks.

PEGMATITES

Pegmatites are extremely coarse-grained rocks which occur as dikes, pipes, or irregular-shaped bodies usually in or near granitic intrusions which have cooled at great depth (Fig. 2). They commonly contain coarse crystals of feldspar, quartz, and mica; some have a variety of valuable and unusual minerals, many of which are peculiar only to pegmatites. There are two kinds of pegmatites: those that form as a result of magmatic action and those that are apparently products of granitization. Isolated pegmatite lenses enclosed in a variety of metamorphic rocks and nowhere near granitic intrusions have been described in Scandanavia and seem

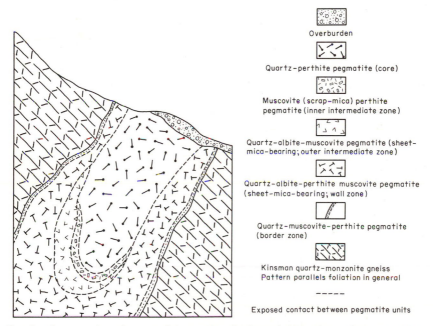

Fig. 2. Cross section of a typical pegmatite, the Pattuck Mica Mine, Alexandria, N.H. (From G. W. Stewart and N. K. Flint, *U.S. Geol. Surv. Prof. Paper 255*, 1954.)

to be the result of granitization. Other pegmatites, including most of those of value, can be related to nearby intrusions and are surely igneous.

An understanding of the origin of igneous pegmatites again requires an examination of the supposed cooling history and solidification of magma. As a granitic magma cools in a chamber some two or more miles below the surface, biotite, orthoclase, and quartz slowly crystallize from the liquid and become consolidated into a coarse-grained granite. The greatest rate of cooling is at the top of the magma chamber, so that a thick layer of granite forms first at this place while deep within the chamber the magma is still liquid. Near the end of the rock-forming stage most of the chamber is filled with the newly crystallized granite. However, many pockets and pore spaces in the central part contain residual juices that are rich in the volatile constituents of the magma.

At this time, two effects become important. In the first place, crystallization of the granite minerals has removed most of the liquid in which the volatile constituents were dissolved. This results in a great quantity of gas dissolved in the small amount of remaining liquid, so that much greater pressures are required to keep the gas from escaping. The second effect is caused by cracks which open toward the surface from the top part of the new granite into the heated country rock. They are a result of contraction or shrinkage caused by continued cooling. These cracks provide openings through which the residual fluids of the magma may travel upward, driven by the abnormally high pressure built up by the concentration of the volatile constituents. It is possible that the cracks might even be formed when the internal pressures build up in excess of the confining weight of overlying rock.

The escaping fluids, cooling more rapidly because they are nearer the surface and surrounded by the cooler country rock, commence crystallizing in the cracks, which may be forced open by the great pressure of the injected fluids. The growth of the crystals proceeds at a rapid rate because of the high content of volatile substances which makes the hot liquid very fluid. Dissolved chemical material needed for the continued growth of crystals can diffuse or migrate easily through the thin liquid. For this reason, crystals in pegmatites can grow to great size and probably in a short time. Single crystals are known which have attained a length of over 40 feet and are several feet in diameter. These are extreme examples, but crystals 2 and 3 feet in diameter are common.

In general, the first minerals formed in a pegmatite are the last minerals to crystallize in the parent igneous rock, that is, potassium feldspar, muscovite mica, and quartz. If deposits of these suffer no further change, they are called *simple pegmatites*. However, later waves of cooler liquids from the magma may bring in a wide variety of chemical substances, of which many have unusual compositions. In such instances, *complex pegmatites* form with a variety of very rare minerals.

A common geologic process which is particularly important in the formation of a pegmatite is *replacement*. This occurs when one mineral gradually appropriates in a piecemeal fashion the space once occupied by another mineral. The high pressures and temperatures which exist where pegmatites form make it possible for the low-viscosity liquids to penetrate the very finest cracks and pores of the country rock. In fact, it is believed that the solid minerals of the rock allow the diffusion of ions from this material even within their atomic layers.

Of course, at the high temperatures of pegmatite formation (between 300 and 550°C) chemical reactions proceed rapidly between the minerals of the country rock and the volatile pegmatite solutions which are in very intimate contact. Simultaneous solution of one mineral and the replacement by another from solution take place in such a way that any volume of rock may have a complete mineralogic change without ever developing voids during the process. Not only is the country rock replaced by pegmatite minerals, but the earliest formed pegmatite minerals are often replaced by other materials brought in by later waves of residual liquids from the magma. In fact, the late-stage pegmatite minerals are commonly the valuable and unusual minerals containing such elements as lithium, phosphorus, tin, tungsten, uranium, and beryllium. The early stage of pegmatite development results in minerals similar to those of the parent granite and may be called the *magmatic stage*. The later waves are considered to be the *hydrothermal stage,* because many of the minerals which form at this time are similar to those found in high-temperature veins.

The hot water which is responsible for the hydrothermal stage comes from the magma at a temperature too high for it to be a liquid and at a pressure too high for it to be a gas. It is said to be in a *supercritical* state, in which condition water has a very low viscosity yet still has the ability to carry much mineral in solution. Those who feel that supercritical water most resembles a gas have labeled pegmatites and high-temperature veins as *pneumatolitic* deposits.

The following criteria should be useful in the recognition and classification of pegmatite deposits:

1. Pegmatites are extremely coarse-grained deposits, generally rich in quartz, mica, and feldspar.

2. Many minerals such as spodumene, amblygonite, and lepidolite (all lithium-bearing) are characteristic of some pegmatites. Other typical minerals are beryl, apatite, topaz, and tourmaline, but these minerals are also found in other high-temperature deposits.

3. Igneous pegmatites occur as dikes, lenses, or small irregular bodies within or near to some granitic igneous intrusion. They rarely attain a size greater than 100 feet in diameter.

4. Metamorphic pegmatites occur as isolated lenses or pockets in regions of intensely metamorphosed rocks where no parent granite intrusion is evident.

CONTACT-METAMORPHIC DEPOSITS

When a magma cools at great depth, the heat which it contains, as well as great quantities of water and other volatile substances, must leave the magma chamber. Because heat always flows from a region of high temperature to a region of low temperature and the volatile solutions must travel from a region of high pressure to a region of low pressure, it is apparent that both the heat and solutions always go by some suitable but often indirect path toward the surface where the temperature and pressure are relatively low.

The rocks surrounding the magma chamber, and particularly those above it, are often greatly affected by this heat and by the solutions which are escaping from the cooling magma. The textural and chemical changes brought about in these rocks are the results of that process we call *contact metamorphism.*[2] The word "contact" is used because the most profound effects are always nearest the contact surface between the igneous rock and the country rock. In order to investigate some of these effects and how they take place, it will be convenient to consider separately the effects of heat alone and then the results of mineral-bearing solu-

---

[2] The term "contact metasomatism" is used by many authors to emphasize the importance of replacement (metasomatism) as the process whereby the metamorphic minerals and ore minerals are implanted.

tions on the different kinds of country rock in the zone of contact metamorphism.

Effects of heat. Heat has the ability to inflict great changes on most rocks, especially when water is present. Of course, no chemical substances are added to or taken from a rock by heat, but chemical elements already present may be recombined or rearranged to form an entirely new suite of minerals. Also, the texture of a rock can be very much altered by a process called *recrystallization* whereby crystal growth increases the grain size such that a coarse-grained metamorphic rock may develop from a fine-grained sediment, often without any change in the mineral content. This is strikingly displayed by the contact metamorphism of a limestone in which the action of heat and a small amount of water allows the simultaneous solution of the smaller grains and growth of the larger grains of calcite until the rock becomes a coarse marble.

Other rocks react differently to the heat of contact metamorphism. A shale behaves in the same way that clay does in a brick kiln. It merely bakes to a hard bricklike mass. Sandstone consists mainly of quartz, which is very resistant to heat and to most chemical activity, and so it is scarcely changed at all by heat. Shaly limestones and other mixed sedimentary rocks are more affected than pure limestones or sandstones and yield complex but characteristic suites of new minerals.

Importance of solutions. Although both heat and solutions always work together, the contact-metamorphic effects attributed to magmatic solutions are responsible for the introduction of ore minerals. Solutions may carry from the magma valuable metallic and nonmetallic substances, which are deposited in the country rock in the zone of contact metamorphism. Copper, lead, zinc, tungsten, tin, and many other substances of use to man occur in deposits such as these.

Much silica (silicon dioxide) and many other nonmetallic substances are commonly introduced into the contact zone, causing significant changes in the mineralogy and texture of the affected rocks.

The silica behaves in three ways: (1) It may precipitate directly in the pores of the rocks as a quartz cement. In this way sandstones can be changed to quartzites. (2) It may replace the affected rock. As the solutions carrying the silica dissolve minerals of the country rock, quartz is simultaneously precipitated in the vacated space. Replacement is common not only by silica but by other substances dissolved in the solutions from the magma. (3) The silica in solution can react with chemical substances in the country rock to form silicate minerals.

With the introduction of iron and magnesium into an impure limestone, it changes to a complex rock containing coarsely crystalline silicate minerals (such as garnet, mica, and others) along with hematite, calcite, and perhaps an assortment of ore minerals. Such a contact-metamorphic rock is called *skarn,* a name also used for ore deposits of this nature. Several other kinds of metamorphic rocks are formed in the contact zone, but limestones are so easily and extensively changed and make such suitable hosts for the deposition of ore minerals that skarns lead all others in metal-bearing potential. Any discussion of ore deposits formed by contact metamorphism, therefore, generally centers on limestones.

The importance of fluids in contact metamorphism lies not only in their function of transmitting mineral matter but also as a means of transporting heat from the magma. Naturally, the more extensive contact alterations occur where the most heat and the most solutions encounter the most vulnerable rocks. A rock which allows the easy percolation of these hot fluids may be heated to higher temperatures more quickly and to greater distances from the contact than dense rock through which the solutions cannot flow. All rocks are poor conductors of heat, and dense rock can be heated only by conduction; whereas the permeable rocks are also heated by the solutions which flow through them.

A good analogy is a steam-heating system which has a radiator connected by pipes to coils in the fire box of a furnace. If the system contained no water, heat would gradually move along the empty pipes by conduction, but the radiator may never become very warm before the coils in the furnace melt. However, heat is quickly and efficiently moved through the system if it contains water that is free to circulate. Thus, mineral deposition and rock alteration are commonly greatest in rocks bordering permeable beds, along bedding planes, joints, faults, and other rock openings which may carry hot solutions *upwards* from the contact zone. Figure 3 shows the relationship of contact-metamorphic ore deposits to rock types, rock attitudes, and distances from the contact zone.

Around igneous masses formed by granitization, contact-metamorphic effects are generally severe and widespread. Metallic deposits that occur in such contact zones are believed to be concentrations of the metals originally present in very low percentages in the sedimentary rocks. They were somehow mobilized by the heat and streamed out beyond the region of most intense granitization.

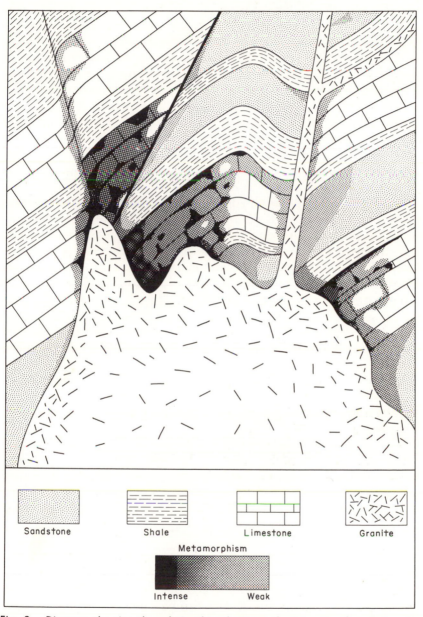

Fig. 3. Diagram showing the relationships between the intensity of contact meta-morphism and the factors of rock type and structure of the country rock.

Pore water in the sedimentary rocks as well as chemically combined water driven from the clay minerals by heat might have moved as hydrothermal solutions which transported and deposited the ore substances and caused alteration effects.

A feature common to all contact-metamorphic deposits is zoning. Bands of most intense metamorphism are nearest the contact, and these grade outward to zones in which only minor changes have occurred. Certain ore minerals are characteristic of the inner zones, whereas others are only found far from the contact. No one is sure of the reasons for such distributions, but certainly temperature plays a major role, in conjunction with such factors as solubility of the metallic compounds, their heats of formation, their volatility, and other chemical and physical properties.

In summary, the combined effects of heat and solutions acting on the country rock in the zone of contact metamorphism immediately adjacent to an igneous intrusion result in changes which depend upon the temperature, the pressure, the nature and attitude of the country rock, as well as the nature of the solutions. Contact-metamorphic deposits have the following characteristics which enable one to identify them:

1. Each one is located adjacent to some coarse-grained igneous rock and rarely extends more than a few thousand feet from the contact.

2. The rock surrounding the deposit is usually highly altered.

3. The ore minerals are not confined to cracks or other openings but seem to have formed from solutions that have soaked into the rock.

4. Most contact-metamorphic deposits occur in highly altered limestones and commonly contain such characteristic minerals as garnet, diopside, tremolite, wollastonite, and a brown mica called phlogopite.

5. Many deposits are discontinuous, irregularly shaped and may show no obvious connection to the contact zone.

HYDROTHERMAL DEPOSITS

Hydrothermal deposits are formed by hot-water mineral-bearing solutions, which in most instances can be shown to have risen from below and to have originated from a magmatic source. Such fluids may also be of metamorphic origin, liberated during granitization. There have also been some proposals that, even without water, gases or diffusion streams of metallic ions may account for the transportation of ore materials from a magmatic source or from a mass of granitized sedimentary rocks. Most evidence, however, points to water as the agent of transportation; and the nearly universal genetic association of hydrothermal deposits and nearby igneous intrusions is inescapable. We should, therefore, direct our attention to the origin of magmatic hydrothermal solutions and mineral deposits formed by them.

Such solutions commonly form sheetlike mineral deposits called veins where deposition has occurred in a crack, fault, bedding plane, or other two-dimensional opening. However, hydrothermal mineralization may take place in any pre-existing rock openings, and, where conditions are suitable, replacement and alteration of the country rock can occur.

Three necessary conditions must be satisfied before hydrothermal deposits can form: (1) An intrusive magma or other source must supply the mineral-bearing water solutions which travel towards the surface under the influence of a pressure gradient. (2) Openings in the country rock must be available through which these solutions may pass. And (3) some chemical or physical mechanism must cause the deposition of the mineral substances from the water solutions.

There is a vast amount of speculation in the science about all three of the above points. We are not sure of the source of the so-called hydrothermal minerals, nor can we be certain they are always transported and deposited by water solutions. Openings have not even seemed necessary for the deposition of some ore bodies. Chemical and physical laws must govern the precipitation of minerals by these mineralizing fluids, but which laws controlled any single occurrence is a question often left unanswered.

In the following pages some of the major points of the magmatic hydrothermal theory are considered.

Hydrothermal solutions. Magmas contain considerable amounts of volatile material, most of which is water. No one knows just what percentage of water occurs in a magma, but indirect evidence points to a range between 3% and 8% by weight. Laboratory studies with melts at high temperatures and pressures indicate that more than 10% of water in a magma would be very unlikely except at exceedingly great depths. The average igneous rock contains 1.15% water in hydrous minerals, and lavas are known with up to 4%. A volcanic glass called pitchstone has a water content which in some cases exceeds 10%, but much of this might be a contamination which entered the lava before it hardened. Studies of volcanic gases and contact-metamorphic effects around intrusive igneous rocks show that granitic magmas have more water and other volatile substances in them than the basic magmas.

So that we may make certain calculations, let us assume a water content of 5% for a granitic magma. This amount can easily be dissolved at pressures equivalent to about 2 miles' depth and at temperatures of as much as 900°C.[3] Consider what happens when such a magma crystallizes slowly at depth as the volatile matter escapes. How much water is given off, and what are some of its effects? For a small granite stock with a volume equivalent to a cube 500 meters on an edge (about 0.3 mile) computations show that a little over 4,000,000,000 gallons of water will be liberated by the magma as it crystallizes. If most of this water were to be funneled to the surface through a few fissures in the roof of the stock, it is likely that any mineral in solution would deposit as vein material in these rock openings. Even if the solutions were extremely dilute, a vast amount of mineral matter could be transported and deposited by such a quantity of water. For example, with a gold concentration of but 0.0001%, 413,000 ounces could deposit, or about $14,500,000 worth. Such an amount of gold precipitated in a few fissures would be a very rich ore deposit. More soluble metals, such as silver, copper, and zinc, could easily be present in higher percentages.

Rock openings. The openings through which hydrothermal solutions can travel are of many kinds and related to different stages of the history of the rock in which they occur. Those openings that are intrinsic and characteristic of a rock are formed by the time a sedimentary rock is consolidated, when an igneous rock is first solidified, or during the regional stresses in which a metamorphic rock is made. They might be considered the *inherent openings*. Openings that are caused by later events, such as igneous intrusion, solution, faulting, or folding, are called *secondary openings*. It is often of great value to know the origin and extent of the rock openings in which hydrothermal deposits occur.

The size and continuity of the openings are also of considerable interest. Continuous openings such as bedding planes and faults can act as major channels for the long-distance movement of solutions, whereas small restricted openings may allow only a limited

---

[3] R. W. Goranson, The solubility of water in granitic magmas, *Am. Journal Sci.*, 5th Ser., Vol. 22, pp. 481–502, 1931.

access of the solutions in a "soaking" fashion. The major channel-ways may be considered the principal arteries for the flow of the hot-water solutions, and the rock pores and other smaller openings are the capillaries which branch off the main avenues of flow. Hydrothermal deposits can form wherever the solutions are able to flow, sometimes in the large cracks or main arteries, sometimes in the smaller openings, and sometimes in both.

INHERENT OPENINGS. The effectiveness of rock openings for transmitting ore-bearing solutions depends not only upon the amount of open space in a rock but also on the nature of the openings. Two separate and measurable properties of rocks must be considered. The first of these is called *porosity,* which can be defined as the percentage of the total rock volume which is open or void space. This property tells nothing about the size or nature of the openings. The other property, called *permeability,* is a measure of the ease with which a rock allows the flow of some liquid through it. Commonly rocks with high porosity also have high permeability, but the reverse is true in some instances. A shale, for example, sometimes may have rather high porosity, but the openings are so tiny that once they are filled with water the capillary attraction of the walls on the liquid is so strong that there will be no flow except when the water is driven by an extremely high pressure gradient. Conseqently, the shale would have a low permeability. On the other hand, a dense limestone may have a low porosity, but nearly all of the pore space consists of bedding planes and joint cracks through which water flows easily. It then has a high permeability. Poorly cemented sandstones and conglomerates have, in general, both high porosity and high permeability, particularly if all of the fragmental grains are of the same size. Of course, the kind and amount of cementation of such sediments has a great effect on both properties.

Igneous and metamorphic rocks may also be permeable enough to allow the flow of ore-bearing solutions. Joints and contraction cracks which develop in igneous rocks as they cool make excellent channel-ways. Lava flows commonly have spongy porous tops caused by the escaping gas bubbles which are trapped as they rise to the cooler, nearly solidified lava near the surface. When other lava flows or sediments bury such flows, horizons of very high permeability are incorporated into the rock sequence. Usual features of many regional metamorphic rocks are slaty and schistose cleavage which develop by the recrystallization and alignment of flaky minerals. Such metamorphic planes have much the same influence as sedimentary bedding planes on the permeability of the rocks. Figure 4 shows some openings common to sedimentary rocks.

It should be emphasized that the property of permeability is a directional feature in most rocks. For example, a slate will readily allow the flow of water in a direction parallel to the slaty cleavage, but it is nearly impermeable to flow at right angles to this. The same is true of a sedimentary rock with well-developed bedding planes.

SECONDARY ROCK OPENINGS. Many of the most important major channels occur as fractures and folds in the rocks of the earth's crust. For this reason, the prospector looking for mineral veins usually confines his efforts to areas which are now or once were mountainous regions. Here rocks have been severely folded and faulted, and numerous igneous intrusions may have supplied the necessary hydrothermal solutions. In many instances faults and folds are formed by the same forces that caused the intrusion of a magma. In any event, the structures must be present at the right place and at the right time to serve as channels for the ore-bearing solutions and as sites of mineral deposition. (Fig. 5.)

FAULTS. The so-called gravity fault is an effect of tensional forces and is a fissure along which the overhanging block has dropped relative to the other one. (See Fig. 6.) The amount of movement along this plane may be only a few feet or as much as several thousand feet. Rarely is the fault one simple plane but may be a series of semiparallel planes making a band of fractures often many tens of feet wide. This band, called a *shear zone,* is commonly a greatly fractured mass of rock with a very high permeability. Many faults are marked by zones of rock so badly macerated that they have the consistency of clay. Such material has been given the name *gouge* by miners. Gouge commonly has a low permeability and limits the amounts of fluids that can travel along a fault plane. Brittle sedimentary rocks tend to make clean breaks or porous shattered zones, whereas shales and many igneous rocks make "tight" fractures with much gouge and low permeability. The greater the amount of movement along the fault plane, the more tendency there is for gouge to form. For this reason, so-called minor faults with slight displacement are often the most favorable locations for ore veins. Faults which formed near to the surface are generally more "open," that is, with a higher permeability, than those which occur at great depth.

Thrust faults are the result of compressional forces. These are

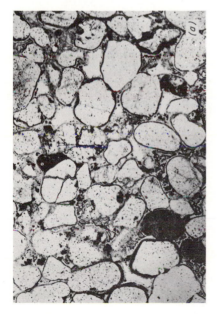

Fig. 4. Some examples of inherent openings in rocks through which solutions can move. (a) Photomicrograph showing nature of intergrain openings in a sandstone. (b) Bedding planes are common openings in sedimentary rocks. (Photograph of exposure of Edwards formation courtesy of Humble Oil and Refining Co.) (c) Most rocks have joints. This is a view of a horizontal bedding plane of the Dakota sandstone showing right-angle joints.

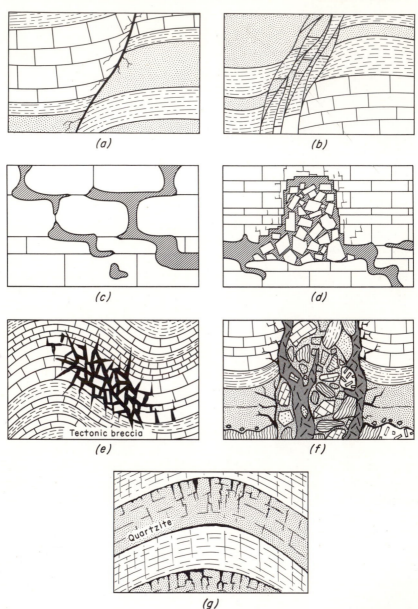

Fig. 5. Examples of secondary openings that form in rocks and make favored channels through which solutions can move. (a) Fault, (b) shear zone, (c) solution caverns in limestone, (d) collapse breccia in limestone, (e) fractures in folded brittle beds, (f) explosion volcanic vent, (g) tension openings in folded beds.

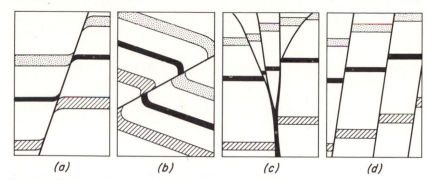

Fig. 6. Kinds of faults and fault systems. (a) Gravity, (b) thrust, (c) branching, (d) multiple.

faults in which the overhanging side has moved upward relative to the other side. The planes of thrust faults commonly have gentle dips, but some are known with dip angles up to nearly 90°. In contrast, most gravity faults are steeply dipping. Many thrust faults contain gouge and are "tighter" (with lower permeability) than gravity faults because of the compressional nature of their origin. Consequently, they are not so important as sites of deposition by hydrothermal solutions, although a few valuable deposits have formed in such structures.

Multiple faults or fault systems are very common. In many mining regions the ore occurs in subparallel groups of faults which have all been mineralized at the same time. Where one fault intersects another, a zone of greatly disturbed rock with a high permeability is created. Faults commonly branch upward like the limbs on a tree. Many small faults at the surface may be traced to one single "feeder" at depth. Within certain igneous intrusions, and rarely in other massive rocks, vertical zones of intense shattering have formed with networks of cracks so fine that an unfractured piece the size of a baseball can seldon be found. Where the rock surrounding such shattered material is dense and impermeable, great vertical "pipes" for the flow of hydrothermal solutions are created by the network of fractures. Such a shattered zone with a nearly circular cross section is called a *stockwork* and may be a very important locale for the formation of a disseminated hydrothermal ore deposit. Many of the important mining districts of the world are classified as stockworks. One of the largest is the tin-bearing stockwork at Altenberg, Germany, which has a diameter of over 3000 feet.

FOLDS. The compressional forces which cause rocks to be crumpled into anticlines and synclines also create suitable openings for the transmission of ore fluids. These openings develop for many reasons, one of the most common of which is tension created at the crest of folds. Tension may open up cracks which extend the length of synclinal and anticlinal axes and are particularly evident in the more brittle rocks. Even the gentle warping of brittle beds near the surface may result in a band of shattered breccia which easily allows the flow of ore solutions (Fig. 5).

An anticlinal fold is like an architectural arch in many respects. If there are some very rigid and competent beds layered between others which are soft and pliable, much of the load of overlying rock above the crest of the anticline is supported by the limbs of the folded competent beds. This naturally causes a decrease in the pressure just under the crest of each folded competent layer. No large openings are created, but the lowering of pressure results in the migration of fluids to these points. Plastic deformation of the incompetent beds generally causes them to thicken at the crests and troughs of the folds, and shear faults are formed in all units nearly parallel to the bedding. Where hydrothermal fluids are drawn into such structures, replacement of the thickened incompetent beds and the filling of shear faults may result in saddle-shaped bodies of ore called *saddle reefs.*

Folded rock strata can act like inverted troughs under which rising ore solutions may be funneled to the surface. When an impermeable bed such as a shale lies above a very permeable sandstone or limestone in a flat-lying sequence, solutions coming up from below encounter the shale and cannot penetrate it but spread out in a vast sheet below it. Should this rock sequence be folded into an anticline, the rising solutions would collect under the arched-up shale and continue to move upward in the direction opposite to the plunge of the fold axis (Fig. 7). The fluids, thus forced to flow in a restricted channel, must deposit their mineral load so that it is concentrated in a limited volume of rock, with the possible formation of an ore deposit. In this instance no new rock openings are created, but the structure has caused a more concentrated flow.

Deposition from hydrothermal solutions. In previous pages the source of the ore-bearing solutions has been considered, and some of the ways have been discussed in which these solutions are able to travel upward through the rocks of the crust of the earth. The reasons why minerals precipitate from these solutions in ore-bearing veins must now be explored. Such considerations lead into a high-

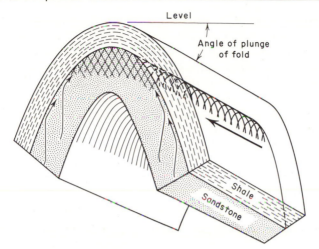

Fig. 7. Solutions collect and move upward under the arch of impervious beds in plunging anticline.

temperature–high-pressure realm of chemistry where the reactions are complex and have not been thoroughly studied. However, some of the chemical principles which are believed to guide mineral precipitation from hydrothermal solutions may be examined without treating with specific reactions and details.

EFFECT OF TEMPERATURE. The effects of temperature on the solubility of mineral matter in water are familiar to everyone. For example, it is common knowledge that more sugar or salt will dissolve in hot water than in cold. Now suppose that a certain amount of water at 100°C will dissolve 100 grams of salt, after which no more salt will dissolve. This, then, is a *saturated* solution. If the same amount of water at 20°C becomes saturated with 94 grams of salt in solution, what will happen if the first saturated solution is cooled down to 20°? Naturally, 6 grams of salt will precipitate. It is likely that hydrothermal solutions are saturated with many different dissolved minerals, and they start to the surface at high temperatures (as high as 500°C). As they travel away from their hot source, the cooling action of the surrounding rocks may easily cause the precipitation of some of the mineral matter dissolved in saturation amounts. The minerals precipitating in this manner form a coating of crystals on the walls of the rock openings through which the solutions are moving.

Zoning of mineral deposits concentrically away from some apparent heat source is commonly observed and is surely an effect of temperature. Rarely, however, are the characteristic hydro-thermal minerals of each zone arranged outwardly in order of increasing solubility as one would expect. Consequently, some other temperature-controlled property must be responsible for the zone deposition. The crystallization of any mineral results in the liberation of a definite and measurable quantity of heat which is called its *heat of formation*. There seems to be a closer relation-ship of mineral zoning to heats of formation of the hydrothermal minerals than there is to their solubilities. There are several un-explained exceptions even to this property, however, and it is prob-able that different effects operate simultaneously; no one can explain all zoned mineral deposits.

A widely used classification of hydrothermal deposits is based on their temperatures of formation. A threefold division was first proposed by Lindgren, who divided the gradational range of hydro-thermal deposits into: *hypothermal* (300 to 500°C), *mesothermal* (200 to 300°C), and *epithermal* (50 to 200°C). There are marked differences in the mineral contents, structures, and textures of the three vein types by which we can distinguish them. Their tem-peratures have been determined by laboratory investigations on the stabilities of characteristic minerals.

The following table contrasts the hypothermal and epithermal veins. Mesothermal deposits share some properties and minerals of the two extremes.

HYPOTHERMAL VS. EPITHERMAL MINERAL DEPOSITS

Hypothermal	Epithermal
1. Usually in ancient metamorphic or igneous rocks	1. Usually in younger rocks of all kinds
2. Textures like those of coarse igneous rocks; interlocking grains, crude band-ing, low porosity	2. Many open cavities, vugs, fractures, incrustations, good banding, comb structure
3. Molybdenum, tin, and tungsten ores are common	3. Mercury and antimony ores are common
4. Characteristic ore minerals: magnetite, specular hematite, cassiterite, molyb-denite, scheelite, rutile	4. Characteristic ore minerals: cinna-bar, stibnite, native copper, native silver, many arsenic and antimony metallic sulfides.
5. Characteristic gangue minerals; amphi-boles, pyroxenes, garnet, tourmaline, topaz, arsenopyrite, pyrrhotite	5. Characteristic gangue minerals: barite, adularia, alunite, zeolites, chalcedony, marcasite

EFFECTS OF PRESSURE. It is well known that some of the gases which are liberated by magmas cause acid solutions when dissolved in water. Hydrogen chloride forms the potent hydrochloric acid; carbon dioxide makes the weak carbonic acid, and other magmatic gases such as hydrogen fluoride and sulfur dioxide also make strong acids. Because the dissolving power of such solutions is directly related to the amount of gas in solution, it would be well to consider the effects of pressure on the solubility of gas in water. It has been previously established that liquids will hold more gas in solution under conditions of high pressure than at low pressure if the effect of temperature is not considered. Therefore, it is possible to speak of saturated solutions of gas controlled mainly by pressure, in the same way that saturated solutions of mineral matter are a function of temperature. If water is saturated with gas at a high pressure, it will liberate some of that gas should the pressure decrease. Because pressure in the earth increases with depth, hydrothermal solutions saturated with gas at great depth will give off some of the gas when they reach shallower depths.

Many mineral substances can remain in solution only if these waters also contain dissolved gases. Therefore, the escape of gases from some hydrothermal solutions must certainly be accompanied by the precipitation of mineral matter. As an example of this process:

$$\underset{CaCO_3}{\text{Calcite}} + \underset{H_2O}{\text{Water}} + \underset{CO_2}{\text{Carbon dioxide}} \rightleftharpoons \underset{Ca(HCO_3)_2}{\text{Calcium bicarbonate}}$$

The left-hand side of the foregoing equation represents a limestone (calcite) through which is percolating ground water containing dissolved carbon dioxide gas. A water solution of carbon dioxide is known as carbonic acid. Thus, in the above reaction the calcite, acted on by carbonic acid, is changed to soluble calcium bicarbonate, which is then dissolved in the ground water. The amount of calcite that can dissolve depends upon the amount of carbon dioxide in solution, and the quantity of this gas in solution depends upon the temperature and pressure. The higher the pressure on the above situation the more calcite will dissolve. Suppose that the pressure is released from the ground water containing dissolved calcium bicarbonate. Note that the above reaction also has an arrow pointing back to the left. This means that as long as the four substances are together the reaction proceeds in both directions at once, and some calcium bicarbonate is constantly changing back to the three substances listed on the left. A lowering of the pressure allows the carbon dioxide to escape, so that the reaction cannot continue to produce calcium bicarbonate, but calcium bicarbonate continually breaks down by the reverse reaction, precipitating insoluble calcite and liberating more carbon dioxide gas. Of course, after the gas has escaped, this calcite cannot redissolve.

It is easy to visualize solutions rising from deep within the earth, pressure decreasing as they near the surface, the carbon dioxide bubbling out of solution, and calcite crystallizing at the same time. Many hydrothermal veins contain considerable calcite along with the ore minerals.

We should keep in mind, however, the counteracting effect of temperature on the solubility of gases in water. Because more gas dissolves in cold water than in hot, a drop in the pressure on a saturated water-gas solution may not liberate any gas if the temperature also falls sufficiently.

ACIDITY. It has been suggested that hydrothermal solutions, which have their origin in cooling magmas, are likely to be acidic because of the gases dissolved in them. Some of the mineral matter dissolved in these waters can remain in solution only if the water retains its acid character. Should these waters encounter rocks containing calcite or other carbonate minerals, a chemical reaction would take place dissolving the carbonate, but at the same time the acid solutions would be neutralized or become alkaline. When the water is no longer acid, many of the mineral substances which it contains would be precipitated. This is possibly one reason why veins commonly contain more valuable ore minerals where they cut through beds of limestone. Many of the ore minerals in contact-metamorphic limestones may be deposited for the same reason.

Some hydrothermal solutions, such as those that deposit pegmatites, are thought to be alkaline. Certainly most acid fluids would soon become alkaline by reaction with the carbonate and silicate rocks.

METAL EXCHANGE. Chemists long ago established the fact that metals differ widely in their ability to enter into solutions. They speak of this property as the tendency or potential of the metals to become "ionized." It is also a well-known principle that, if a metal in solution meets a compound containing a metal with a greater tendency towards ionization than the one in solution, the first metal will come out of solution and the new metal will take its place. This can easily be demonstrated by the simple experiment of placing iron nails in a solution of copper sulfate ($CuSO_4 \cdot$

$5H_2O$, blue vitriol).   The ionized copper leaves the solution and forms a plate on the iron as a layer of copper metal, while some of the iron, which has a greater tendency towards ionization, goes into solution.   For the same reason, if a hydrothermal solution containing iron comes in contact with calcite ($CaCO_3$), the calcium with a greater ionization tendency takes the place of the iron in solution, and the iron takes the place of the calcium in the mineral. This forms the new mineral, siderite, $FeCO_3$.

Certainly, there are many other chemical principles which can be applied by the geologist to explain the deposition of mineral matter from hydrothermal solutions, but those briefly mentioned above are the most easily understood and offer logical explanations for many hydrothermal deposits.

Some characteristics of hydrothermal deposits which set them off from other types of mineral deposits are as follows:

1. All hydrothermal deposits show that they were introduced into the rocks which contain them.

2. Veins and other hydrothermal deposits are related to structural features through which solutions moved.

3. Replacement hydrothermal deposits are often scattered patches of ore minerals with no apparent connection to major solution channels.

4. Cavity-filling hydrothermal deposits can be recognized by their shapes which conform to the natural rock openings they have filled.

5. Many characteristic minerals such as those mentioned in the table on page 38 are typical of hydrothermal deposits.

SEDIMENTARY DEPOSITS

Ore deposits which fall under this classification are really no more than unusual sedimentary rocks which through some deviation of the "normal" pattern of sedimentary history contain one or more minerals in concentrations rich enough to be exploited. Like any sedimentary rock, these deposits have followed the usual geologic events: (1) sedimentary material provided by the weathering and erosion of earlier rock, (2) transportation of this material in some manner to a sedimentary basin, (3) deposition and accumulation of the material in the basin, and (4) consolidation into a sedimentary rock.   However, somewhere during the above stages of development a variation from the normal conditions or an unusual source for the material has resulted in a rock containing a concentration of a valuable metallic or nonmetallic mineral. Several deposits of iron and manganese are of this type, as are many of the rich nonmetallic mineral deposits of the world.

Many famous and important ore deposits are today the subjects of heated debate as to whether or not they are sedimentary. These include the Rand gold deposits; Blind River–Algoma uranium district; Tri-State zinc and lead; Kupferschiefer copper, lead, zinc ores; and the Colorado Plateau uranium deposits, all of which are discussed in later pages.

Advocates of granitization rely on sedimentary-rock sources for all metals in lode deposits, but concentration by metamorphic process is necessary. C. L. Knight[4] has proposed a sedimentary source bed, one with higher than normal percentages of one or more metals occurring probably as sulfides. When the bed is heated excessively by an igneous intrusion or by granitization fluids the original sulfides are forced to migrate from the source bed. Where the migration of the sulfides has been slight, a definite stratigraphic relationship of the ore bodies is evident. The theory offers a nice explanation why all the ore bodies in some districts are confined to certain beds and also why certain geographic areas are characterized by deposits of one particular metal.

The placer deposit, discussed in detail below, is the result of sedimentary processes of streams and the sorting and transporting ability of marine waves and currents. Most other sedimentary deposits result from circumstances peculiar to each occurrence, and examples are treated in the chapters devoted to specific metals or nonmetals.

An ore deposit which is sedimentary in origin may be recognized by the following set of criteria:

1. It will commonly have many of the features peculiar to the ordinary sedimentary rocks, such as fossils, bedding planes, ripple marks, etc.

2. It will normally occur as a bedded sedimentary rock interlayered between rocks of undoubted sedimentary origin.

3. The minerals may be those which form in a weathering environment and are particularly resistant to the agents of weathering and transportation or they are chemical precipitates normal to surface waters.

---

[4] C. L. Knight, Ore genesis—the source bed concept, *Econ. Geol.,* Vol. 52, pp. 808–818, 1957.

4. It need not be associated with any evidence of igneous or metamorphic activity, so necessary for the formation of most other deposits discussed so far.

5. It is often a deposit of great geographical extent.

6. It may be of recent geologic origin, thus occurring at or very near the surface.

Placer deposits. Mineral deposits in which certain heavy and valuable minerals have been concentrated by the selective transportation action of running water are called *placers.* A placer concentration can only occur when the valuable material is a substance which has a higher density[5] than the worthless substance with which it occurs. Also, the valuable material must be of such a chemical and physical nature that it can resist the chemical processes of weathering and the mechanical wearing action of stream transportation. A few of the minerals which meet these requirements and are often found in placers are garnet, magnetite, corundum, hematite, rutile, gold, diamond, cassiterite (tin oxide), platinum, and monazite (a phosphate of many rare metals). Of these, only the last six are recovered in commercial amounts from placers.

The placer gold deposit is perhaps the one kind of ore deposit known to most people. When a prospector goes into the mountains to pan gold on some mountain stream, he is searching for a rich placer deposit of gold-bearing gravel. It was a placer deposit that was discovered at Sutter's sawmill in 1847 and set off the great gold rush to California. The same was true for the Klondike gold rush of 1896. Much of the gold produced today still comes from placer deposits, but giant dredges and hydraulic methods have replaced the pan and the sluice in larger operations. Gold placers rich enough for small-scale mining might be considered the "poor man's ore deposit." They are easier to find than other ore deposits; they consist of easily mined loose gravels and sands which generally occur near the surface, and the gold can be simply and cheaply extracted with a minimum of equipment. Besides this, a miner can bring out his whole summer's production in the saddle bags of his horse.

For a placer deposit to form there must be a primary deposit that contains gold or other valuable substances that may be concentrated. The primary deposit should be exposed to weathering

---

[5] Density may be defined as the weight of a substance divided by its volume. One cubic centimeter of quartz weighs 2.65 grams; hence, its density is 2.65 grams per cubic centimeter. Pure gold has a density of 19.33 grams per cubic centimeter, etc.

on a slope where erosion can remove the debris and carry it to a stream channel. Weathering merely frees the valuable mineral grains from the rock matrix. When the products of weathering, valuable substance included, are washed into a rapidly flowing stream, the actual sorting and concentration of the valuable mineral occurs. The nature of these processes may be explained by the principles governing the transportation of sedimentary material by running water.

All the sediment which is being moved by a stream under a certain set of conditions is called the *load* of that stream (Fig. 8). A stream can roll large particles along the bottom by the simple "push" provided by the flowing water. Particles moving in this way are considered the *traction load* of a stream. The *suspension load* consists of smaller particles that are able to remain up off the stream bed, suspended by the turbulent forces of the stream flow. Other particles intermediate in size bound along, sometimes in suspension and sometimes as part of the traction load. These are called the *saltation load* of a stream. A fourth method of transportation of mineral matter is by *solution*.

It is also important to know why, at a certain rate of flow of the water, the stream is able to move some particles while others remain at rest on the stream bed. Surely, *size* of the particles is a controlling factor. A stream that can roll pebbles along the bottom may not be able to move large boulders. However, if the velocity of the stream flow increases sufficiently, even the boulders may be moved. *Shape* also plays an important role. A well-rounded shape enhances movement in the traction load, whereas a flaky particle is more easily transported by suspension than a well-rounded particle of the same size. The property of *density* is prob-

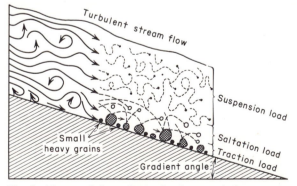

Fig. 8. Nature of the sediment load in a turbulent stream.

ably the most important factor. Two particles may be of the same size and shape, but one may have a density twice that of the other and consequently weigh twice as much. The stream may have the ability to move the lighter particle but not the one with the higher density. Therefore, a stream flowing with a certain speed would have the same ability to move a large particle with low density as it would to move a small particle with a high density. There are many other physical principles which affect the transportation of sedimentary materials by streams, but the above should be sufficient to explain the formation of placer deposits.

Now let us see what happens to the debris from the weathering of some primary mineral deposit that has entered a rapidly moving mountain stream.

The first effect of the running water is to sort the debris into the different stream loads. When the stream velocity is high enough, sand, silt, and clay become the suspension and saltation loads which rapidly move downstream. These are followed by the slower traction load which includes the small but heavy grains of ore minerals along with the large but light pebbles and cobbles. Thus, an initial concentration of the ore minerals is achieved, because much of the worthless part of the original debris has been removed.

The valuable concentrate in the traction load must come to rest somewhere to form the accumulation which is a placer deposit. Slowing of the stream flow will cause most of the original traction load to come to rest. Thus, an accumulation takes place in the stream's course where the water for one reason or another is slowed down slightly. Too much of a decrease in velocity is not desirable, as this would cause a simultaneous deposition of the saltation and suspension load along with the traction load and undo all the work of separation so far accomplished.

One would look for a traction-load accumulation in the stream's course: (1) immediately upstream from any obstruction which tends to pond the stream flow, (2) just downstream from where a tributary enters its slower moving master stream, (3) upstream from an outcrop of steeply dipping resistant beds which form low ridges of bedrock along the stream bottom, (4) in the gravel bars which form in the slack waters at the inner part of curves in the stream course, and (5) in pot holes and plunge pools which form at the base of waterfalls and rapids (see Fig. 9). Naturally, deposits such as these which have formed in streams that are no longer in existence are more difficult to recognize. In all cases, however, the placer deposit has the nature of a sandy gravel, consisting of coarse sand,

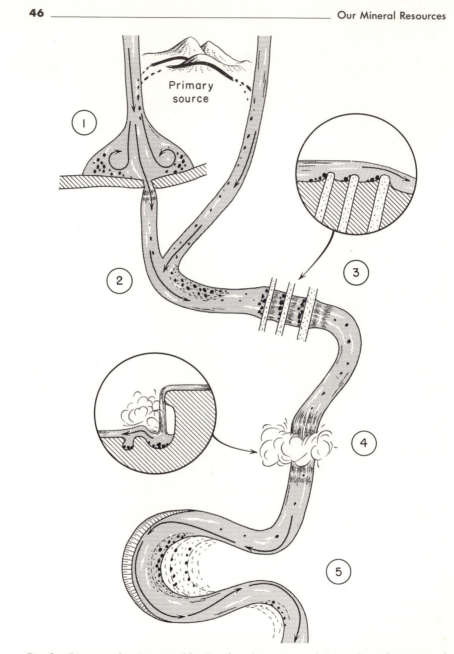

Fig. 9. Diagram showing possible sites for placer accumulations along the course of a stream: (1) behind an obstruction that dams the stream, (2) downstream from a tributary mouth, (3) behind ridges formed by dipping resistant beds, (4) in plunge pools and pot holes, (5) in slack waters along inner shores of meander loops.

pebbles, and particles as large as boulders. The ore particles are usually smaller than most of the other material in the deposit, although occasionally larger grains and nuggets of ore minerals can be found.

It should be mentioned that under certain conditions placer deposits may be formed by the action of waves on beaches. Even eroding winds in dry regions may cause a slight degree of concentration of heavy ore minerals by blowing away lighter dust material.

A search for and recognition of stream placer deposits should be guided by the following physical characteristics:

1. The texture of placer deposits is usually that of a sandy gravel, yet the ore minerals are commonly small particles of heavy minerals included in the gravel.

2. Many rich placers lie on bedrock at the base of a gravel, the small ore minerals having worked their way through the coarse material.

3. Because most primary ores, which must supply the ore minerals to placer deposits, occur in areas of highly contorted rocks (usually mountainous regions), valuable placers will generally form, if at all, in streams draining such areas.

4. Meandering sluggish streams are not able to transport coarse material and heavy grains, so that the search for modern placers should be confined to rapidly flowing waters. "Dry washes," through which seasonal drainage moves, are equally satisfactory.

5. Stream situations that cause sudden velocity decreases in the flow are sites of gravel accumulations which may be valuable placers.

DEPOSITS FORMED BY SECONDARY ENRICHMENT

Deposits of secondary enrichment are those in which weathering and sedimentary agencies have concentrated valuable substances from any kind of previously existing mineral deposit, which could be an igneous rock, a sedimentary rock, a vein deposit, a pegmatite, or even a previously formed secondary-enrichment deposit. The parent deposit may or may not have been rich enough to be considered an ore. Many secondary-enrichment deposits are the result of recent geological events and are found near or at the surface. Others have formed from weathering processes in the ancient past and occur near unconformities buried below a sequence of sedimentary rocks. Many secondary-enrichment deposits are actually soils which develop on some parent material. These form only where the products of weathering have undergone little or no transportation.

There are really only two effects of weathering which can produce a deposit of secondary enrichment; (1) *residual concentration,* removal by solution of material which is of no value, leaving behind a concentration of a valuable substance, and (2) *supergene enrichment,* a selective solution and removal of the valuable substance which is carried downward and redeposited in a concentrated state a short distance from the parent material.

Residual concentrations. These mineral concentrations are deposits where worthless mineral matter is chemically changed to a more soluble state and removed. The valuable material is left behind and concentrated because it is impervious to the attacks of chemical weathering or is altered to a new mineral which will not go into solution or be further affected by chemical weathering. Naturally, mechanical erosion of the products of weathering must be at a minimum.

A number of conditions are necessary for the formation of residual concentration deposits: (1) A primary mineral deposit must be available which contains a valuable substance that is resistant to chemical attack. (2) The deposit must be exposed to the effects of weathering. (3) The outcrop surface should have low relief so that gravity and running water cannot remove the insoluble products of weathering. (4) An adequate supply of moisture should be present, at least periodically, to carry off in solution the more soluble products of weathering. (5) The above conditions must remain in effect for long periods of time. Chemical weathering is a very slow process, so that a million years or more may be necessary to achieve sufficient concentration of the insoluble and valuable substance.

To illustrate the process of residual concentration a simple experiment is instructive. Use a long glass tube with a plug of filter paper at the bottom. The tube should be filled with rock salt containing 5% of lead bird shot well mixed into the salt. This mixture represents the primary deposit which contains ore minerals (the bird shot) in amounts too small to be economically recovered. Now cause "chemical weathering" to act on the make-believe deposit by allowing water to slowly drip into the top of the tube and gradually seep out at the bottom. The bird shot is, of course, unaffected by the water, but the salt is dissolved and carried out through the filter. If the process should be stopped when half of the salt is gone, the "ore minerals" would then make up a little over 9.5% of the mixture. When three-fourths of the salt is removed, the "ore deposit" contains the valuable substance in the amount of 17.4%; the "chemical weathering" could be continued until the "deposit" is 100% pure ore minerals. The formation of such deposits in nature (Fig. 10) requires the removal by solution of great quantities of mineral substance, but examples of all degrees of concentration occur between a slight increase in the ore-mineral percentage and the nearly complete removal of all the worthless substance from the original deposit. Because most of the mineral matter removed by weathering (such as feldspar, clay, calcite, and even quartz) is only very slightly soluble, long periods of time are needed for any appreciable results.

A surficial layer of weathered and enriched ore occurs on many deposits of lead, copper, tin, mercury, and other metals that can remain in an insoluble chemical state during the process of weathering. In some instances the enriched material is the only substance that is concentrated enough to be worked as ore.

In humid, tropical and subtropical areas of the world chemical weathering has created large and very valuable residual deposits of high-grade iron and aluminum ores. These two metals, when they are combined with oxygen and hydrogen as *hydroxides,* are the least-soluble common metals. They are left behind after all the other rock constituents have been gradually dissolved away.

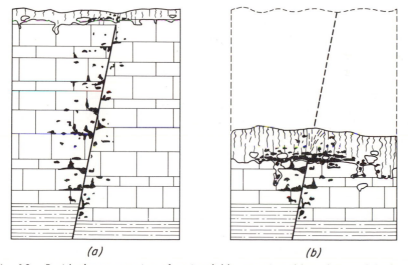

(a)   *(b)*

Fig. 10.   Residual concentration of an insoluble ore mineral by solution of the limestone in which it occurred. (a) Primary mineral deposit in limestone before weathering, (b) secondary concentration in soil above weathered limestone.

Ground waters are charged with oxygen, carbon dioxide, and even organic acids from the decay of vegetation. The dissolving power of these reagents is so great that even silicate minerals are decomposed and the silica carried away in solution. Such conditions may be contrasted to the environment of chemical weathering in a temperate climate where the aluminum becomes combined with silica to form one of the many clay minerals. Under temperate climatic conditions no further weathering can remove the silica from the aluminum, and the end product is a deposit of clay. In tropical weathering even clay will be further attacked and the silica removed, leaving an end product rich in aluminum hydroxide (bauxite). This is the main ore substance of aluminum. Such a weathering process so extreme that silica can be removed has been named *lateritization,* and a deposit formed by such conditions is called a *laterite.* Laterite deposits require all the conditions necessary for the formation of any residual concentration, but in addition there must be a hot humid climate. The parent material, in varying stages of weathering, lies directly below the deposit, unless lateritization has affected it all.

Deposits of residual concentration have the following properties which make possible their identification:

1. They were all formed on comparatively level surfaces where mechanical erosion was not able to remove the products of weathering.

2. They are commonly stained a deep red or brown color because of the iron oxides which are always concentrated in the weathering zone. A large enough deposit of this kind may be a valuable iron ore, or some other valuable substance may remain in commercial amounts with the iron oxide.

3. Deposits of residual concentration are generally very porous or have the consistency of loose soil because of the great quantities of mineral matter which have dissolved.

4. They are nearly always underlain by their parent material. This may be a primary mineral deposit, or in the case of laterites may be any rock containing aluminum- or iron-bearing minerals.

5. Laterite deposits occur in parts of the world which are now or once have been under the influence of a humid, tropical or subtropical climate.

Deposits of supergene enrichment. These are deposits which can best be defined by the etymology of the word *supergene.* The prefix "super" means "from above"; "gene" comes from the same roots as the word "genesis" and here has the meaning of "origin." Hence, *supergene* means "origin from above." This refers to the precipitation of ore minerals from ground water soaking down from the surface, and the term is the antonym of *hypogene,* which is used for deposits by hydrothermal solutions rising from below.

Supergene enrichment does not usually create a new ore deposit but rather is a process whereby surface waters dissolve the ore minerals from the weathered zone of an exposed deposit, carry the dissolved material downward, and redeposit it just below the ground-water table. The metal values carried down in this way and added to the metal content of the primary material often make a valuable ore out of what may have been a very lean or entirely worthless deposit. An explanation for this process requires consideration of some of the elementary principles of chemistry discussed in earlier pages.

In order for an ore mineral to be dissolved in the zone of weathering, it must be chemically changed to a soluble state. Most primary ore minerals are simple chemical compounds in which the metals are combined with sulfur to form sulfides. These minerals weather by a series of chemical changes in which oxygen is added to the sulfides and changes them into sulfates, which are nearly always more soluble. The process is aided by the fact that pyrite (FeS_2) is present in almost all primary deposits. Weathering of pyrite results in two soluble forms of iron sulfate and sulfuric acid. All three are potent reagents that attack and dissolve the other primary sulfide ore minerals. The common metallic sulfides which are most easily dissolved in this way are those of iron, copper, silver, and zinc. Many other metallic sulfides are not converted into sulfates by this process (such as cinnabar, HgS), or they form sulfates which are very insoluble in water (such as the sulfate of lead). The ease of solution of the various metals is controlled by the ionization abilities of these metals, and their precipitation in the zone of enrichment is controlled by the same inherent property.

All the above conditions of weathering and solution of ore minerals take place in that part of the ore deposit which is unsaturated by ground-water and where the pore spaces in the rocks are filled with air (Fig. 11). Only slight and occasional moisture is present to accomplish the necessary solution, but plenty of oxygen is available for the conversion of sulfides into sulfates. This is known as the *zone of oxidation*. It is usually capped by a rusty and porous-looking mass of material called a *gossan*. Just as in lateritization, the red to brown iron oxides and hydroxides which

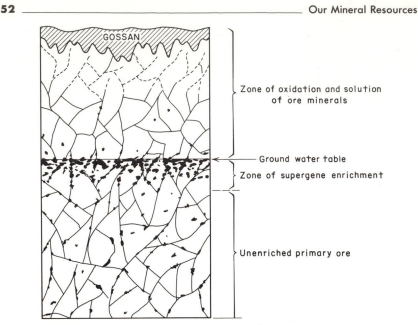

Fig. 11. Diagram showing zones of oxidation and supergene enrichment as they would develop on a deposit of disseminated copper sulfide minerals.

form during weathering are extremely insoluble and are left behind as an accumulation at the surface. The spongy nature is caused by the removal of the more soluble material, leaving voids. Whether or not a zone of enrichment lies below, the outcrop of most ore deposits is capped by gossan, so that the prospector is constantly on the alert for such rusty-looking streaks which may reveal the presence of valuable ore below the surface.

Before any enrichment can take place, the metals, which have dissolved as sulfates in the zone of oxidation, must be reprecipitated in the *zone of supergene enrichment*. This zone lies just below the upper limits of the water-saturated rock or soil, which is known as the *ground-water table*. Below this surface the pores of the primary deposit are entirely filled with water. Here no oxygen is available for further weathering changes, and an entirely different chemical environment exists. This is a zone, therefore, which contains all of the original ore minerals of the deposit essentially unchanged, and it is invaded by solutions containing dissolved metals. In such a situation the previously described chemical principle of *metal exchange* determines what takes place. Iron occurring in primary sulfides seems to be the controlling metal in the reactions of enrichment. Those metals in solution which have an ionization ability lower than iron take its place in pyrite and other sulfides while the iron enters into solution. Copper and silver are the only common metals with lower ionization potentials than iron. Hence, supergene enrichment is rarely observed in deposits of any other metal. As zinc has a greater tendency toward ionization than does iron, all of the zinc which is dissolved in the zone of oxidation is dissipated and forever lost.[6]

To illustrate this enrichment process consider the mineral changes in the zone of supergene enrichment in a primary deposit of copper. Starting with pyrite in the following table the progressive changes to chalcocite indicate a removal of iron and its replacement by copper. Of course, any chalcopyrite or bornite in the primary ore will be acted upon by the solutions to form the next mineral that is richer in copper (Fig. 12). Copper may also replace the metal calcium in calcite or any other metal which is more subject to ionization than itself. Silver has the added advantage of replacing copper, and the chemistry of its supergene enrichment is complicated indeed.

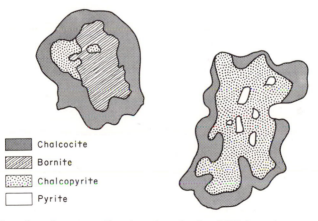

Chalcocite

Bornite

Chalcopyrite

Pyrite

Fig. 12. Drawing of two metallic mineral grains ($\times$ 125) from the supergene zone of the copper deposit at Bingham, Utah, showing mineral relationships of the table on page 54.

---

[6] Zinc may be retained in the zone of oxidation if it combines to form insoluble carbonates and silicates.

MINERAL CHANGES IN THE SUPERGENE ENRICHMENT OF COPPER

Mineral Names	% Iron	% Copper	% Sulfur
pyrite ↓	46.6	—	53.4
chalcopyrite ↓	30.5	34.5	35.0
bornite ↓	11.1	63.3	25.6
covellite ↓	—	66.4	33.6
chalcocite	—	79.8	20.2

There are certain conditions which can prevent supergene enrichment by restricting the solution of the valuable metals. For example, an appreciable amount of calcite in a primary deposit will not allow copper sulfate to be carried down to the ground-water table but will react with it to create the relatively insoluble blue and green copper carbonates that remain in the zone of oxidation. A slight amount of salt in the ground waters acting on a primary silver deposit will immediately cause the formation of the very insoluble silver chloride. Such conditions may, however, permit the accumulation of valuable metals in the zone of oxidation, where they would form deposits of residual concentration.

The process of supergene enrichment does not make a new and separate mineral deposit but merely adds to the value of some earlier-formed metallic ore. Its effects are usually only detected when the primary ore is either test drilled or mined, and assays show a zone of richer metal values. Therefore, no attempt will be made to suggest criteria for the recognition of such enrichment zones in the field.

SELECTED REFERENCES

Adams, F. D. _The Birth and Development of the Geologic Sciences,_ Williams and Wilkens, Baltimore, 1938.

Clarke, F. W. Data of geochemistry, _U.S. Geol. Surv. Bull._ 770, 1924.

Garrels, R. M. Solubility of metal sulphides in dilute vein forming solutions, _Econ. Geol.,_ Vol. 39, pp. 472–483, 1944.

Graton, L. C. Nature of the ore-forming fluid, _Econ. Geol.,_ Vol. 35, Supp. to No. 2, 1940.

McKinstry, H. Mining geology: retrospect and prospect, _Econ. Geol.,_ Vol. 50, pp. 803–813, 1955.

Sullivan, C. J.   Ore and granitization, *Econ. Geol.,* Vol. 43, pp. 471–498, 1948.
Sullivan, C. J.   Heat and temperature in ore formation, *Econ. Geol.,* Vol. 52, pp. 5–24, 1957.

GENERAL REFERENCES

The many excellent textbooks by Bateman, Lindgren, Emmons, McKinstry and others may be consulted for more detailed treatment of many of the topics discussed above.

iron

INTRODUCTION

Iron is the fourth most common element of the earth's crust and the second most common metal. It has been estimated that a little over 5% of the rocks of the outer "shell" of the earth consists of iron. Aluminum, which occurs with an average abundance of slightly over 8%, is the only metal more common than iron. For the most part, the rocks at the weathered surface of the earth contain iron in chemical combination with oxygen and water. Such iron makes its presence very evident by the red, yellow, and brown coloration which it imparts to the rocks. Except for a minor occurrence in Greenland, the only metallic iron found in the natural state is from meteorites, which are of extraterrestrial origin.

Uses for Iron

The modern metallurgy of iron requires great quantities of other metals and nonmetals for the many highly specialized *steels*. Table

1 lists some of these substances and tells what special property is imparted to the steels because of their presence. By varying the "recipe" amounts and by special treatments, the metallurgist may greatly increase the range of properties.

TABLE 1. COMMON ALLOY ELEMENTS AND THEIR EFFECTS ON STEEL

Alloy Element	Character of Steel	Uses
low carbon %	soft and ductile	wire, tubing, boiler plate, rivets, nails
medium carbon %	harder	axles, forgings, and machinery
high carbon %	very hard	chisels, drills, files, dies, machine tools
manganese	tough and abrasion resistance	rock crushing and digging equipment, curved rails
nickel	high elasticity, ductility, resistance to corrosion, hardness	armor plate, structural steel for bridges
chromium	hard, temperature and corrosion resistant, high tensile strength	dies, ball bearings safes, stainless steel
molybdenum	shock resistant, strong, tough, hard	automobile parts, tools, machinery
vanadium	like molybdenum steel, but also fatigue resistant	shafts, axles, and steel springs
tungsten	hard and tough even to dull red heat	"tool steel," high-speed cutting tools, magnets, armor-piercing shells
silicon	high electrical resistance, large magnetic permeability	electrical equipment
copper	much the same as for small amounts of nickel	corrosion-resistant tubes and sheet metal
wrought iron (low carbon with small amount of slag "kneaded" in)	tough, corrosion resistant, malleable	ornamental metal work, pipes for wet ground

* Table based on pages 171–172 of *Minerals in World Affairs* by T. S. Lovering. Copyright, 1943, by Prentice-Hall, Inc., Englewood Cliffs, N.J.

Production

The United States has always been a major producer of iron ore, and, although our richest ores are dwindling rapidly, we were still able to produce about 103,000,000 tons in 1955. This amounted to slightly less than 30% of the total world's production of 366,400,-000 tons for this year. It shows a marked increase over the 75,000,000 tons of iron ore mined in 1940 when the United States produced 35.3% of a world production of about 206,000,000 metric tons. The future looks bright for the continued ore resources available to our mighty iron and steel industry. New methods of utilizing lower grade ores have greatly increased our domestic reserves, and recently discovered high-grade deposits in Brazil, Ontario, Quebec–Labrador, and Venezuela add greatly to world reserves.

GEOLOGY OF IRON ORES

Ore Minerals of Iron

There are scores of minerals in the rocks of the earth which contain some iron, but of these only four are of any importance in the ores of iron. These minerals and the percentage of iron which they contain are: (1) hematite (Fe_2O_3)—70%, (2) magnetite (Fe_3O_4)—72.3%, (3) siderite ($FeCO_3$)—48.2%, and (4) limonite[1] ($FeO(OH) \cdot nH_2O$)—about 60%. Hematite and magnetite are both oxides of iron, siderite is a carbonate, and limonite is an earthy hydroxide of iron. Most ores of iron as they are fed into the blast furnace contain between 35 and 65% iron; it is clear that very little worthless mineral matter is present in the ores, which must be nearly pure masses of ore minerals.

Magmatic Iron Ore Deposits

Most of the very basic igneous rocks contain between 6 and 8% iron. To appreciate the quantity of iron present in a limited amount of magma picture a "small intrusion" of gabbro with a volume equivalent to a cube 500 meters on a side. Gabbro is a rock with an average iron content of 6.84%. If but one-half of the iron in the original magma crystallized as magnetite and if this magnetite were segregated while the magma was crystallizing, an ore deposit containing 195,000,000 tons of magnetite would be created. In this

---

[1] Limonite is not a mineral in the strict sense of the word but a mixture of several iron minerals with an earthy, amorphous hydroxide of iron.

one small deposit there would be enough iron ore to supply American needs for over 1½ years. Of course, rarely in nature is there ever so complete a segregation, but one deposit comes to mind which exceeds the hypothetical example in quantity of ore and closely approaches its content of pure magnetite. This is the great *Kiruna Deposit* of Sweden, one of the most important of Europe.

KIRUNA DISTRICT

This famous mining district is located in the northern part of Sweden 100 miles north of the Arctic Circle. The main deposit consists of a dikelike mass which is 475 feet wide and extends along its outcrop for about 1¾ miles. The ore is nearly pure magnetite, and, since the ore body is more resistant to erosion than the surrounding rocks, its outcrop stands as a ridge some 750 feet above the surface of nearby Lake Luossa Jaidi. (Fig. 13.) Drilling has shown that the ore mass dips to the east only 20° from vertical and extends to a depth of at least 1000 feet. Estimates of the total quantity of ore in this one body are close to one billion tons. Across the lake from the Kiruna dike is the smaller but similar Luossavaara ore body. This ore mass is about 4500 feet long and 130 feet in width. Just east of Luossavaara is the smaller Rektor ore body that has been mined since 1942 mainly for its high phosphorus content.

All three ore masses are thought to be magmatic in their origin,

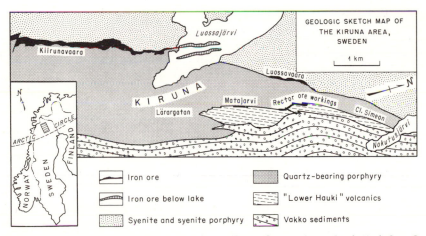

Fig. 13. The Kiruna iron district, Sweden. (From Geijer, *Sver. Geol. Und.* Ser. C, No. 514, Stockholm, 1950.)

although the mechanics of their formation are not very well understood. The most widely accepted theory of the origin of the Kiruna deposits is as follows: In the distant geologic past the region was covered with several thick and extensive lava flows upon which was deposited a sequence of normal marine sediments. These rocks were steeply tilted by some early period of mountain building. In the process, the rocks were severely metamorphosed. Later, an intrusion of some basic rock, such as a gabbro, seemingly took place at a depth still far below the present surface. As this magma cooled, the processes of crystallization in some manner resulted in a concentration of iron oxides or hydroxides in the last remaining "juices" of the magma. The chemical nature of this liquid can only be surmised, and the history of chemical and physical changes in the magma which allowed the concentration of iron in the late fluids is not known. In any event, before the magma was completely crystallized, compressive forces resulted in a filter-pressing action that forced the residual liquids upward into rock openings available in the overlying rocks. Apparently, the contact surface between two of the old lava flows easily admitted the liquids, which crystallized as the thick magnetite dikes of the Kiruna and Luossavaara ore bodies. Another sheetlike opening between the quartz-bearing porphyry and the overlying "Hauki" volcanics received the liquids that formed the smaller Rektor ore body. Subsequent erosion has uncovered the deposits, and recent glacial activity has formed the lakes and deposited extensive glacial debris on the present surface.

All of the ores from this district are contaminated with phosphorus from the mineral, apatite $(Ca_5F,OH(PO_4)_3)$. The Rektor ores contain up to 7% phosphorus which is extracted for use in fertilizers. Because of this impurity in the iron ore special metallurgical techniques are required which are economic only because of the very high iron content of the ore (between 56 and 71%). Sweden's production from this district and the smaller Gallivare district, about 50 miles south of Kiruna, was over 16,000,000 tons in 1951. Most of the ore is exported to Germany.

Sedimentary Iron Ore Deposits

These deposits of iron were formed by the deposition of solid and dissolved sedimentary material in marine basins of accumulation. Sedimentary ores are perhaps the most common and widely used in the world, but some iron-rich sedimentary rocks become ore only after suitable processes of secondary enrichment.

CLINTON IRON ORES

The hematite-rich Clinton formation was laid down in the seas which invaded North America some 320 million years ago in that period of the geologic past called the Silurian. The ores were deposited as lenses included between alternating beds of marine sandstone and shale, at intervals when neither mud nor sand were accumulating very rapidly. The deposition of iron oxide seems to have been controlled by a prior, or concurrent, accumulation of calcite-rich material. In favorable locations on the sea bottom, beds of calcite mud were deposited in what was probably rather shallow agitated water. Calcite formed the shells of tiny molluscan creatures which accumulated on the sea bottom. Also, tiny spheres of calcite, generally less than a millimeter in diameter, were developed from what was probably a gelatinous calcite ooze. These have been named *oölites* because of their resemblance to clusters of fish eggs. In some places the oölites have a flattened appearance, giving them a resemblance to flaxseed. For some unknown reason the waters of these Silurian seas must have been well saturated with iron in solution. Apparently, before the calcite mud was buried by later sediments, iron oxide precipitated from the sea water and replaced calcite which was redissolved. Even the shells, oölites, and "flaxseeds" were replaced by hematite. Some recent experiments have shown that when stream waters carrying iron in solution enter into a well-aereated shallow marine basin containing calcite, some calcite dissolves and iron oxide precipitates. No one knows why the streams carried so much iron to the seas at this particular time of the geologic past.

The Clinton beds have a wide geographic extent and appear in outcrops that extend from Pennsylvania and New York along the Appalachian Mountain chain into north-central Alabama. However, only in a few places along this strip is the ore rich enough or favorably located to warrant mining. An outcrop of the Clinton beds in eastern Wisconsin along the western shore of Lake Michigan contains iron ore of the same type. The region near Birmingham, Alabama, has extensive reserves of excellent ore close to coal deposits of a grade suitable for the production of coke. In addition, the nearby occurrence of limestone, necessary for blast-furnace flux, has made Birmingham one of the major steel-producing cities of the world. Its steel supplies most of the markets of the southeastern states and furnished nearly all of the steel to support the armament program of the South during the Civil War.

There are two kinds of ore mined in the Birmingham district; the "hard ore," which is the primary sedimentary rock containing about 36% iron with considerable calcite and a little clay; and the "soft ore," which is a residual concentration of iron found close to the surface. The "soft ore" is richer, with about 50 to 60% iron. The hard ore would seem to be of rather low grade, but the main impurity is calcite, a substance usually added to the blast-furnace charge. Whereas most iron ores must be mixed with limestone when they are fed into the blast furnace, the Clinton ores already contain sufficient calcite impurities to act as a flux during the smelting process. The mines of the Birmingham district operate in four separate seams of iron ore in the Clinton rocks, which dip to the southeast at an angle about 30°. The thickest of the four ore horizons, appropriately named the "Big Seam," is 16 to 40 feet thick, of which 7 to 12 feet is minable. Nearly all of the ore is taken by underground operations. The available ore reserves have been estimated at about 1.9 billion tons. A large thrust fault, which formed as a result of the same compressional forces responsible for the crumpling of the Appalachian mountain belt, has carried a slice of old rocks, including the Clinton beds, northwestward and left them close to the younger coal-bearing beds of Carboniferous age. This fortunate geologic circumstance brought iron ore near to valuable coal beds which otherwise would have lain many miles to the west.

The Lorraine ores of Central Europe and the Cleveland Hills deposits of England are sedimentary iron ores similar to the Clinton beds.

BOG IRON ORES

Bog or swamp deposits are not of great importance for the production of iron, but they do point to some interesting geologic principles explaining the precipitation and accumulation of iron in fresh-water lakes and swamps. Because some of these deposits are forming at the present time in the northern glacial lakes, it is possible to observe the processes of iron accumulation and to speak with some certainty about the chemical principles which govern the precipitation.

Streams emptying into the swamps carry iron in solution as an iron bicarbonate ($Fe(HCO_3)_2$), iron sulfate ($FeSO_4$) such as forms from the weathering of pyrite, and iron humates, which result from the solution of iron by humic acids generated in soils by rotting vegetation. Some of the iron is deposited along the stream course, but when the stream waters flow into the still, stagnant depths of a swamp, many complex reactions take place. There are bacteria in the waters of the swamp, and these, like all plants, require carbon dioxide for their life processes. This gas is soon exhausted from the water, so that the tiny creatures are forced to extract the carbon dioxide from the iron bicarbonates. This leaves iron hydroxide ($Fe(OH)_3$) which, being insoluble, precipitates and accumulates on the bottom of the swamp. The iron sulfate reacts with calcite, in solution in all such waters, and with the oxygen near the surface of the swamp waters. The product of these reactions is a film of limonite (iron hydroxide) which floats as a scum on the surface of the swamp. Such scum often has an irridescent sheen like a film of oil and has given many people the false hopes of discovering an oil seep where swampy conditions occur. However, wave action soon breaks up the limonite scum into tiny rafts which eventually sink to the bottom and add to the accumulation of iron. The humates of iron are also precipitated as limonite by a more complex chemical alteration in which oxygen and the small bacteria play an important role. Other less common minerals of iron along with quartz, calcite, and clay make up the mineral content of most of these deposits.

Iron Ores of Secondary Enrichment

This classification of iron ores includes the largest and most important deposits in the world. For the most part, they have formed by residual concentration of many kinds of iron-bearing rocks of a grade too low to be utilized as ores. Most iron ore deposits have a surficial zone of enrichment, but the deposits discussed below are only rich enough to be considered ore where this enrichment has taken place. In all cases where such iron concentration has occurred, the primary iron-bearing rock lies just below the ore. The primary rock in this case is known as the *protore*. A protore is any rock which is not rich enough to be utilized as an ore, but from which ore has been formed by some process of secondary enrichment.

LAKE SUPERIOR IRON ORES

The most important ore deposits in the United States are located around the western and southern shores of Lake Superior within the borders of Minnesota, Wisconsin, and Michigan.   There are seven principal mining districts (Fig. 14), the largest and most productive of which is the Mesabi Range.   The iron ores were first discovered in 1844 by a surveying party on what is now the Marquette Range.   Except for test shipments there was no great development of the ore until 1855 when the ship canal at the Sault was completed.   By 1857 a railroad had been laid between the mines and the lake, and water shipping to the eastern industrial cities rapidly increased.   By 1870 the Marquette Range was shipping nearly a million tons of ore each year.   The Menominee, Gogebic, and Vermillion ranges were discovered in the early 1880's, and these with the Marquette are sometimes known as the Old Ranges.   The Mesabi Range was completely covered by glacial

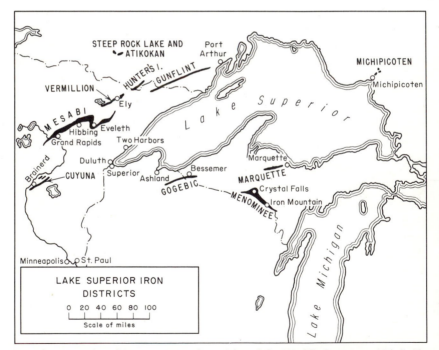

Fig. 14.   Reproduced by permission from *Atlas of the World's Resources,* Vol. II, *The Mineral Resources of the World,* by William Van Royen and Oliver Bowles. Copyright, 1952, by Prentice-Hall, Inc., and published for the University of Maryland.

drift, and its discovery shortly before 1892 was a direct result of magnetic surveys which guided a drilling program through the glacial mantle. The deflection of the magnetic compass needle by such quantities of iron proved a valuable prospecting guide in the search for more and better iron ore. The remaining ranges were discovered in the early twentieth century also as a result of magnetic methods of prospecting.

The brief historical sketch above has gone back in time a little over a century. Let us now make a big jump to the distant past about 1½ billion years ago when the protores of these iron deposits were beginning to accumulate in an ancient marine basin. There must have been a wide, rather shallow sea covering much of the area and enclosed by low-lying and barren land. The climate was probably much warmer than it is today, and the seas contained none of the creatures which populate our present-day oceans. We shall first turn our attention to the land areas, for it was here that the iron was present in the rocks. It had previously been a region of great mountains with highly crumpled rocks and many large intrusions of granite, but the processes of weathering had since eroded it to a gentle featureless surface. There was probably some life on the land surface in the form of primitive plants, such as algae, moss, and lichens. Otherwise, conditions of soil formation and weathering were probably much as they are now. Because of the low relief, chemical weathering was the main form of disintegration of the exposed rocks, and it resulted in solutions containing some calcite, iron bicarbonate, silica, and salts of potassium, sodium, and magnesium. Most of the silica and much of the iron were probably carried in the form of colloidal particles.[2] It is believed that organic acids and colloids not only played a part in the rock weathering but also were necessary as a protective coating on each iron and silica colloidal particle to keep it from joining particles of opposite electric charge. Such coagulation would have built up particles of such size that they would have settled out. When the colloid-bearing solutions entered into the salty marine waters, the many electrically charged particles called *ions* (which result from the solution of any salt) were attracted to the oppositely charged colloidal particles despite the organic coating. The particles were

---

[2] These are extremely small solid particles which have identical electric charges that prevent their sticking together to make larger particles. Colloids also have a coating of water molecules. They do not settle to the bottom because of the continuous agitation of the water molecules around them.

neutralized, thus allowing them to clump together to form watery, gelatinous blobs that quickly settled to the bottom. These blobs were iron hydroxide, iron silicate, silica, and calcium and magnesium carbonate. Often a layering resulted in what eventually became beds of these substances after compaction squeezed the water from the ooze. Some clay, sand, silt, and other sedimentary particles were incorporated. A layer of sand lay beneath the iron-bearing beds, and a thick accumulation of mud covered them. Much of the above geologic history is speculative, but it is a picture which is mostly consistent with facts available.

At some later time the southern part of the area was covered by a vast sequence of lava flows. The region was later intruded by the Duluth gabbro mass, which in most areas assumed the character of a sill of great thickness. This intrusion cut off the eastern extent of the deposits and greatly altered what is now the eastern third of the Mesabi iron beds. Subsequently, folding formed a broad gentle syncline of these rocks with an axis located somewhere near the center of what is now Lake Superior.

A somewhat different history of sedimentation, tighter and more complex folding, some faulting and later intrusive granite dikes, make the geology of the Michigan iron ranges considerably more complex. Even after all the events related so far, there still had not been created any iron ore, but merely a banded cherty iron-bearing rock with a granular texture. This rock, called *taconite,* is the protore from which the later processes of secondary enrichment were able to concentrate the iron.

The formation of the iron ores from these ancient sedimentary rocks was brought about by solutions that caused the oxidation of the iron-bearing minerals of the taconite and the leaching of the silica and carbonate. Most geologist believe that ground waters soaking downward through the taconite were responsible for the formation of the ore bodies.[3] Such leaching near the surface might be classified as weathering, in fact, even lateritic weathering, but in many districts the solutions percolated downward to great depths and leached the taconite far below the surface.

The beginning of these enrichment processes must have started in the very distant past when the rocks were first exposed by the erosion of overlying beds. Geologic evidence indicates that the seas

---

[3] Gruner has proposed that hydrothermal solutions could have performed the leaching and that these solutions were derived from the Duluth gabbro intrusion that must have once lain above all the ore districts.

covered this area only a few times in subsequent ages, and the thin deposits of sediments which they left were quickly eroded away. Consequently, the iron ores could have been forming over a span of about one billion years, or at least intermittently during this time. It is fairly certain that most of the enrichment occurred long before the advent of the glaciers which once covered this region and that the deposits were undoubtedly of much greater extent before the sheets of glacial ice plowed up and removed much of the ore. In the Mesabi Range solutions removed the silica only locally along the 100-mile outcrop at places where slight buckling had created in the brittle taconite a tensional shatter zone with high permeability. In other districts (Fig. 15) the solutions worked their way downward along troughs of higher permeability and were commonly confined to the upper surfaces of inclined dikes. In some places fault zones allowed the penetration of the weathering solutions. These downward-moving solutions dissolved silica and created irregular pockets and bands of enriched ore well beneath the surface. For this reason, the Marquette, Gogebic, Crystal Falls, and Menominee ranges are worked by underground mining methods. In most cases the enriched ore is generally a soft concentration of nearly pure hematite, but locally the ores have been metamorphosed to a hard shiny black flaky form called *specular hematite*. These must have been ores that were enriched very early and were involved in the compressional movements which created the folding and faulting of the area.

THE MESABI RANGE

A description of each of the many Lake Superior deposits is not of great value in a book of this scope, but some further mention should be made of the great Mesabi Range, from which most of the ore of the district is produced. As has already been stated, the taconite or iron formation crops out as a 100 mile band 1 to 3 miles wide which extends from near Pokegama Lake northeastward to Birch Lake. Except for a zigzag at its center near Virginia, Minnesota, the trace of the outcrop is nearly a straight line on the map. The beds of taconite are over 600 feet thick, trend about 73° east of north, and dip to the south about 9°. They are part of the north limb of the major syncline of the region. The ore occurs in small pockets along the outcrop particularly where fracturing has allowed the penetration of solutions. Most of the ore is produced from the western two-thirds of the range, for the eastern part has been so

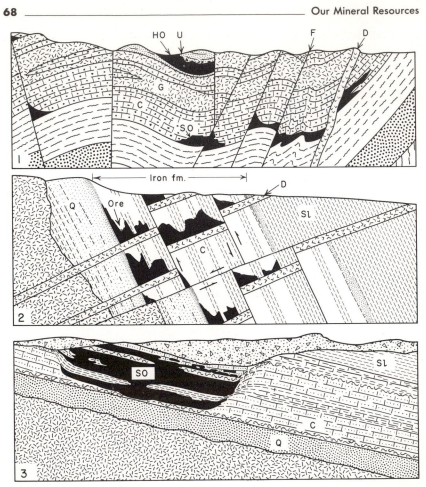

Fig. 15. Diagrammatic sections in Lake Superior iron districts showing occurrence of ore. (1) Marquette range, Michigan. HO, hard crystalline ore; SO, soft ore; F, fault; D, dike; G, gabbro sill; C, siliceous iron carbonate rock of iron formation from which the iron ore is formed by oxidation of the carbonate and leaching of the silica. The solutions that did this work were guided by structural troughs in the formation, created by folding, faulting, or intrusion. (2) Gogebic range, northwestern Michigan–northern Wisconsin. Sl, slate hanging wall; Q, quartzite footwall of iron formation; D, dike. (3) Mesabi iron range, northern Minnesota. (Reproduced by permission from *Minerals in World Affairs*, by T. S. Lovering. Copyright 1943, by Prentice-Hall, Inc.)

affected by the intrusion of the Duluth gabbro that little concentration has occurred here. However, the taconite of the eastern part has a high content of magnetite formed by the heat and solutions from the gabbro. Because a process has recently been perfected for the magnetic separation of the magnetite from low-grade taconite, this material is now being mined and utilized as ore. Most of the enriched ore is mined from huge open pits where large electric shovels can easily handle the loose concentrations of hematite. The ore is loaded onto trains which go down into the pits. The trains carry it to the docks in Duluth where it is loaded onto ships that take it to the mills in the eastern Great Lakes area and in Chicago. Over 35 millions tons of ore are shipped annually from the Mesabi Range. The reserves of high-grade ore are probably down to about 500 million tons, but there still are great reserves of minable lower grade material.

IRON ORES OF LABRADOR AND NORTHERN QUEBEC

This recently discovered district has good possibilities of becoming one of the major sources of iron ore in the world. Already many hundreds of million tons of high-grade ore have been proved, and there is every likelihood that much more will be found by continued exploration. A railroad has been completed and mines are producing.

The vast arctic reaches of northeastern Canada are mostly underlain by a complex of ancient Archaen gneisses. Lying unconformably on these rocks is a sequence of Proterozoic strata in an elongate depression trending roughly northwest-southeast. This depression, called the "Labrador trough" is 600 miles long and reaches a maximum width of about 60 miles (Fig. 16). Most of the ore deposits discovered so far lie along the western side of the trough where the Proterozoic rocks include a number of iron formations of the Lake Superior type.

These formations, the Ruth and the Sokoman, are the protores in which are developed all of the ore bodies. They are underlain by quartzite and are covered with slates, a sequence strikingly like that in most of the Lake Superior ranges. The iron formations are tightly folded parallel to the long direction of the trough and the folds are overturned to the southwest. Thrust faults dipping northeasterly parallel these folds, and basic dikes cut the rocks in many places. A granite batholith has intruded and metamorphosed the Proterozoic rocks in the southwest portion of the trough, and in the northeast the Archean and Proterozoic rocks are separated by a major fault. Of course, none of the rock units or structures is persistent along the whole 600 mile length of the trough.

The iron formations average about 30% iron in oxide and carbonate minerals. Quartz and a variety of silicate minerals com-

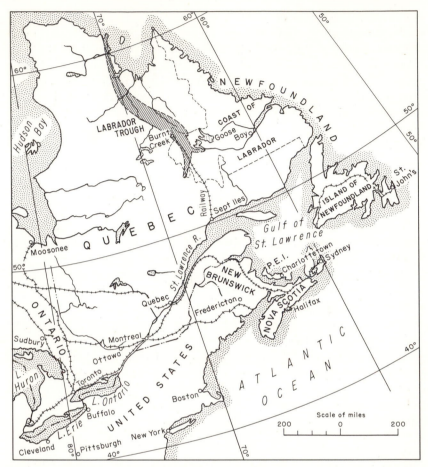

Fig. 16. Map showing the location of the iron deposits of Labrador and Northern Quebec. (After Harrison, *Int. Geol. Cong.* 19th, Algeria C. R. Sec. 10, f. 10, 1953.)

prise the rest of the rock. The ore bodies have been formed at the crests of folds, along faults, and in places along the limbs of the folds where nearly all of the minerals but the iron oxides have been leached from the protore. It is believed that downward-moving ground water was responsible for the formation of ore where it was able to penetrate easily in those localized areas of permeability caused by regional deformation.

The iron formations themselves are so similar to those of the Lake Superior ranges that the same explanations could apply for the manner in which they were deposited in ancient marine environments. The basins of accumulation were probably shallow and partially cut off from the main ocean areas. Conditions were stable for long periods, but there was some alternation of periods of sand and clay deposition with the long intervals of chemical accumulation.

CERRO BOLIVAR, VENEZUELA

Cerro Bolivar is the largest and most important of a number of iron ore deposits to be found in Venezuela since 1946. All of these deposits occur in an 80 mile wide belt of quartzite hills that dominate the topography of the nearly uninhabited region just south of the Orinoco River and its tributary the Caroní (Fig. 17). The Spaniards mined iron ore in the area probably 150 years ago, but modern development of the ores started in 1920 when the El Pao deposit was discovered. This iron mine is now owned and operated by the Bethlehem Steel Corporation. In 1946 the United States Steel Corporation started a large-scale exploration program in this part of Venezuela. Many geologic parties searched the jungles, and they were guided and aided by the results of aerial photographic and aerial magnetic surveys of the area. Cerro Bolivar was first examined in April 1947, and several smaller deposits were also disclosed by the search.

Cerro Bolivar is a ridge about 4 miles long and standing 1800

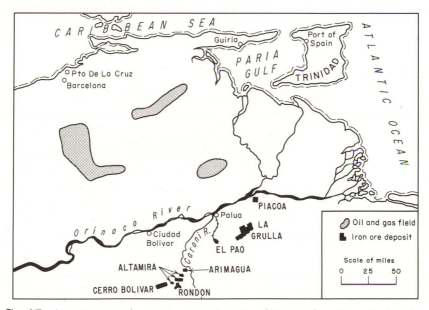

Fig. 17. Location map showing iron ore deposits of Venezuela. (From M. C. Lake, *Eng. and Min. Journ.,* Vol. 151, No. 8, 1950. Copyright McGraw-Hill Publishing Co.)

feet above the surrounding country. The ore bodies lie along the crest of the ridge for nearly its whole length. The width of the ore bodies is variable but reaches a maximum of almost 4000 feet near the center of the ridge. An average ore thickness of 230 feet has been determined, and a maximum of 550 feet was drilled. Over one-half billion tons of high-grade ore have been proved with an average iron content of 63.50%. It is nearly free of sulfur and has very little other objectionable impurities. Great landslides of broken ore are numerous along the flanks of the ridge, and much of the ore from the pits is poorly consolidated. Fine particles screened from the ore will require nodulizing before they can be used as blast-furnace charge.

The ore bodies lie in a syncline that trends about east-west and coincides almost exactly with the top of the ridge (Fig. 18). Cross buckling has divided the syncline into several smaller units some of which pitch east and others west. Folded with the ore beds and lying under them along most of the ridge is a ferruginous quartzite that has up to 40% iron. This bed lies unconformably upon a

		Hole 65	Tunnel E3	Hole 67	Hole 85	Hole 90
Iron	%	60.73	64.16	64.5	61.50	63.17
Phosphorus	%	0.18	0.13	0.127	0.185	0.10
Silica	%	0.83	0.52	0.52	0.63	0.92
Manganese	%	0.11	0.10	0.10	0.08	0.09
Alumina	%	3.41	1.50	1.14	2.06	1.01
Titanium oxide	%	0.19	0.08	0.000	0.21	0.03
Sulfur	%	0.004	0.002	0.054	0.001	0.009
Ignition loss	%	8.36	6.29	6.51	9.16	7.84
Feet of ore		140	670	340+	365+	400

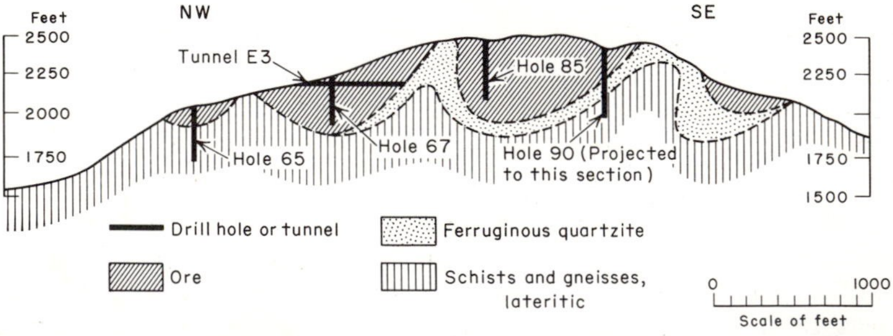

Fig. 18. Cross section through Cerro Bolivar, showing generalized geologic relationships. Exploratory drilling and tunneling are indicated with composite analyses of ore. (From M. C. Lake, Eng. and Min. Journ., Vol. 151, No. 8, 1950. Copyright McGraw-Hill Publishing Co.)

complex of schists and gneisses. Locally the ore lies upon a soft highly weathered unit called the "laterite," and the quartzite may be missing. All of the ridges in the iron belt of Venezuela are due to the presence of folded quartzites that are more resistant to weathering than the other metamorphic rocks. The beds of iron ore are thought to be present as occasional lenses interstratified with the quartzite and also resistant to the action of chemical weathering.

The ore is a mixture of limonite and hematite with some magnetite. It is finely laminated, but intense weathering has broken down much of it to a sandlike consistency. The limonite is a product of this weathering.

The geologic origin of the ore bodies can only be guessed at until more facts are available. It has been suggested that the iron-rich lenses in the quartzite were deposited in marine conditions as chemical precipitates or as biochemical precipitates of ferric hydroxide deposited by bacteria which used carbon dioxide from dissolved ferrous carbonate.

The ores can be easily mined by the open-pit method, but the problem of transporting them to the steel mills of North America is indeed formidable. Railroads have been laid out, dock facilities constructed at the company town of Puerto Ordaz, and channels dredged and maintained in the Orinoco River. Cerro Bolivar lies 91 miles from the dock area at the mouth of the Caroní and 274 miles by a proposed rail route to a tidewater dock at Barcelona. The first ore shipment was in January 1954, and the yearly production by the end of 1956 had reached 8,000,000 tons.

STEEP ROCK LAKE, ONTARIO

This great new iron-ore district lies 150 miles north of Duluth, Minnesota, and about 3 miles north of the village of Atikokan on the Canadian National Railroad (Fig. 19). Despite its remote setting, the area has been known and studied by geologists for over 60 years. The geology of the Steep Rock region provides important clues to the understanding of the Precambrian history of the Canadian Shield. Early reports on the area suggested the possibility of iron ores, but it wasn't until 1930 that Jule Cross, a Canadian geologist, first proved that ore was present under the water and bottom muds of Steep Rock Lake. However, it took him until 1937 to raise sufficient financial backing, and by 1938 operations had started. It was first necessary to drain the water from all but the westernmost of the four arms of the M-shaped lake.

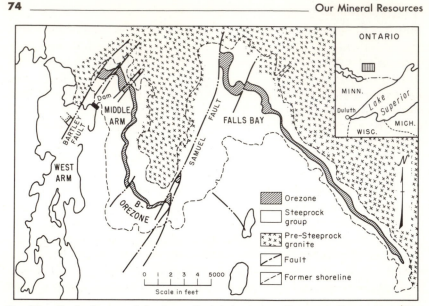

Fig. 19. Geologic map of the Steep Rock Lake area. (From map by A. W. Jolliffe, *Econ. Geol.*, Vol. 50, p. 376, 1955.)

This great engineering feat and other development projects were sufficiently completed so that ore production was started in 1945. About 2.3 million tons of ore with an average iron content of 57.57% was taken from the Hogarth open pit during 1955. Other mines are now going into production, and latest estimates of the reserves are between 1 and 2 billion tons. This is truly a great new addition to the sadly depleted Lake Superior iron ranges. It appears to be genetically more related to lateritic deposits now found in tropical regions.

The ore bodies are part of a sedimentary sequence called the Steeprock group which lies on the eroded surface of an older granite. A metamorphosed limestone locally underlain by conglomerate is the oldest formation of the group. The ore-bearing unit called the "ore zone" rests on the eroded surface of the limestone. Above the ore zone is a thick unit called the "ash rock" that consists of layers of tuff and coarser volcanic debris. Sedimentary rocks, bedded volcanic fragments, and basalt flows make up the youngest formation of the group. The Algoman granite has intruded the Steeprock group and now crops out a few miles to the northwest.

Throughout the area the rocks of the Steeprock group have a strike generally to the northwest, and they dip steeply to the southwest. Many faults with a northeasterly trend cut the area, and their strike corresponds to the foliation direction in some of the metamorphic rocks.   Some sharp flexures, in which the ore zone is markedly thinned, seem to be also related to the same compressive forces from the northwest that caused the faulting and foliation.   The great Samuel fault, with a horizontal displacement of over 2 miles, has truncated the north end of the ore zone that extends the length of Southeast Arm of Steep Rock Lake.   The continuation of the ore zone west of the fault is along the Middle Arm of the lake.   This segment is similarly truncated by the Bartley fault, which apparently has had a horizontal movement of nearly a mile.   Drilling has shown that west of the Bartley fault the continuation of the ore zone is under the still flooded West Arm of the lake.

The ore zone (Fig. 20) consists of two units that have been traced for a total of 4 miles of strike length.   Lying on the erosion surface of the limestone formation is the manganese-bearing "waste rock" with a variable thickness and iron content too low for the material to be mined.   Directly above this is the ore itself, a bed which exceeds 300 feet in thickness in the A-ore body and extends to depths over 2000 feet, as shown by drilling.   The high iron content occurs mostly in the mineral goethite, but some hematite and pyrite are also present.   The ore is soft, porous, and fragmental and contains some masses of kaolinite clay and gibbsite $(Al(OH)_3)$ that bear a strong resemblance to bauxite.

Already many theories have been proposed for the origin of the iron ores of Steep Rock Lake, and surely many more will be heard as further mining and exploration disclose new facts.   The following sequence of events as suggested by A. W. Jolliffe seems to best explain the geology revealed so far, though it leaves some problems still unanswered:

1. A thick unit of dolomitic limestone containing some iron and manganese was deposited on the level eroded surface of the underlying granite.

2. The seas retreated exposing the limestone to a long period of possible lateritic weathering that resulted in a mantle of soil rich in iron and manganese.

3. The area was resubmerged and the soil mantle was reworked by waves and currents.

4. Iron-rich material in vast quantities was carried into the basin and deposited chemically as limonite.   The source of this material was probably the residual soils that were still exposed to weathering on the land.

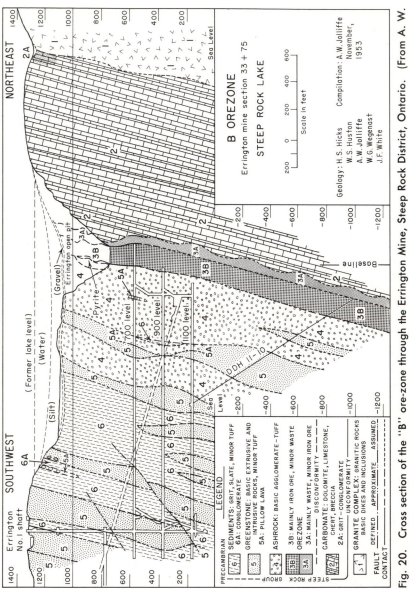

Fig. 20. Cross section of the "B" ore-zone through the Errington Mine, Steep Rock District, Ontario. (From A. W. Jolliffe, Econ. Geol., Vol. 50, p. 384, 1955.)

5. Periods of more normal sedimentation accounted for the few beds of clayey and cherty sediments, some of which weathered during occasional emergences to produce the lateritic material that resembles bauxite.

6. This was all ended by a period of volcanic eruptions during which were formed the thick deposits of "ash rock", volcanic fragments, and basalt flows.

7. The Algoman granite intruded the area, at which time the Steeprock group was tilted to a near-vertical position. Hydrothermal solutions at about 150°C were given off by the granite and caused the crystallization of the limonite to goethite and hematite. Enough sulfur was introduced by these solutions to form the pyrite that is so abundant locally.

8. Faulting caused the horizontal displacement of the ore bodies.

SELECTED REFERENCES

Auger, P. E. The stratigraphy and structure of the northern Labrador trough, Ungava, New Quebec, *Can. Min. and Met. Bull.,* Vol. 47, No. 508, pp. 529–532, 1954.

Bartley, M. W. Steep Rock Iron Mine, *Structural geology of Canadian ore deposits,* Can. Inst. of Min. and Met., pp. 419–421, 1948.

Bateman, A. M. The formation of late magmatic oxide ores, *Econ. Geol.,* Vol. 46, pp. 404–426, 1951.

Burchard, E. F. *Birmingham District, Alabama,* XVI Int. Geol. Cong. Guidebook 2, Washington, D. C., 1933.

Convey, J. (and others). Iron ores in Canada; a symposium (with discussion), *Can. Min. and Met. Bull.,* Vol. 48, No. 516, pp. 213–226, 1955.

Geijer, P. Igneous rocks and iron ores of Kirunavaara, Luossavaara, and Tuolluvaara, *Econ. Geol.,* Vol. 5, pp. 699–718, 1910.

Geijer, Per. The Rektor ore body at Kiruna (in English), *Sveriges geol. undersokning,* Ser. C, No. 514, Arsbok 43, Stockholm, 1950.

Geijer, Per, and N. H. Magnusson. The iron ores of Sweden, *Int. Geol. Cong.* 19th, Algeria, Symposium fer. t.2, pp. 477–499, maps under separate cover, 1952.

Gruner, J. W. *Mineralogy and geology of the Mesabi Range,* Office of the Commissioner of the Iron Range Resources and Rehabilitation, St. Paul, 1946.

Harrison, J. M. Iron formations of Ungava Peninsula, Canada, *Int. Geol. Cong.* 19th, Algeria C. R., Sec. 10, f. 10, pp. 19–33, 1953.

Hotchkiss, W. O. *Lake Superior Region,* XVI Int. Geol. Cong. Guidebook 27, Washington, D. C., 1933.

James, H. L. Sedimentary facies of iron-formation, *Econ. Geol.,* Vol. 49, pp. 235–293, 1954.

Jolliffe, A. W. Geology and Iron ores of Steep Rock Lake, *Econ. Geol.,* Vol. 50, No. 4, pp. 373–398, 1955.

Lake, Mack C. Cerro Bolivar—U. S. Steel's new iron ore bonanza, *Eng. and Min. Journ.,* Vol. 151, No. 8, pp. 72–83, 1950.

Thoenen, J. R., and A. H. Reed, Jr. The future of Birmingham red iron ore, Jefferson County, Alabama, *U. S. Bur. Mines Rept. Invest.* 4988, pp. 1–19, 1953.

Van Hise, C. R., and C. K. Leith. *Geology of the Lake Superior Region,* U. S. Geol. Surv. Mon. 52, 1911.

chapter 3 | aluminum

Uses for Aluminum

This light, strong, and durable metal and its alloys have found a significant and ever-widening place in our present economy. Only a few of the more than 4000 different applications can be mentioned here, but they will serve to show what a versatile metal aluminum[1] is. Naturally, its light weight has made it particularly useful in all forms of aircraft construction. Its resistance to corrosion and its pleasant appearance have led to an increasing utilization in the building industry, where aluminum is used for window sashes, awnings, siding, roofing, and even exterior facing panels for large buildings. Kitchen ware, toys, small boats, and automobile and train parts are all familiar uses of aluminum. Less familiar, but equally important, are the applications of its many

---

[1] The name "aluminium" is used for this metal almost everywhere but in the United States.

alloys for a variety of machine parts where special qualities are required. Powdered-aluminum pigment is used in great quantities by the paint industry, which also employs several of the chemical compounds of aluminum as blue and yellow pigments. Aluminum alum has hundreds of industrial uses, from its application as a dye mordant for fabrics to its utilization in the tanning of leather. In the electrical industry, aluminum has become a serious competitor of copper for many purposes. Aluminum oxide (alumina) is widely used as an abrasive. Nearly 30% of the ore produced is marketed as alumina.

Production of Aluminum Ores

During the peak of the airplane production in World War II the United States mined a record 7,000,000 tons of bauxite a year. The principal American deposits in Arkansas were worked for only about 1,743,344 long tons of bauxite in 1956 and imports supplied most of the ore needed. In the same year, Surinam was the leading country with 3,427,539 tons, then Jamaica with 3,141,330 tons. British Guiana produced 2,481,000 tons. The leading European countries were Hungary and France with 879,000 and 1,442,655 tons of bauxite respectively. The Russian production was estimated at 1,083,000 tons. In 1956 the Western Hemisphere produced over 70% of the world's supply of bauxite. Just recently a large bauxite deposit was discovered on Cape York Peninsula, Queensland, Australia, which was reported to contain many hundreds of millions of tons of ore. Venezuela also had several large new deposits reported in the Guayana region and the Nuria region of Bolivar. However, the latter is in mountainous terrain some distance from water transportation, a situation which will probably delay its exploitation.

GEOLOGY OF ALUMINUM

Ore Minerals

Bauxite, the ore substance from which aluminum is recovered, was named from the village of Les Baux near Arles, France, where it was first described. Bauxite is not a pure mineral substance, but a claylike mixture of several minerals which are all hydrous oxides of aluminum. The earthy mass is commonly stained red or brown by iron oxide impurities and frequently contains pea-size concretionary masses, called *pisolites* (Fig. 21). The mineral *cryolite* (Na_3AlF_6) is a necessary substance for the refining of aluminum. It is a white waxy-appearing mineral which is mined and shipped principally from the village of Ivigtut, Greenland. At the present time, synthetic cryolite is most widely used because of the cost and scarcity of the natural mineral. Clays and other aluminum-rich minerals may someday be ores of aluminum.

Occurrence of Aluminum Ores

Bauxite is formed by only one geologic process, lateritization. Like iron oxides, the aluminum oxides and hydroxides are among the most insoluble major chemical substances in rocks. Feldspars, clays, and micas are rich in aluminum, but in these minerals it is in chemical combination with silica and other substances. The intense weathering of rocks containing these minerals (where they are exposed in flat terrains under hot humid climatic conditions) dissolves the silica, leaving a residue of bauxite and any iron oxide that is present. Many igneous, sedimentary, and metamorphic rocks are suitable for the formation of lateritic bauxite deposits. A low iron content of the original rock results in a bauxite deposit

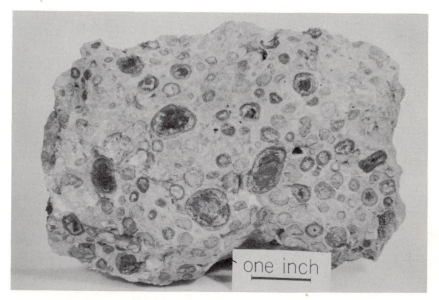

Fig. 21. Bauxite showing pisolites.

with a desirable low content of iron oxide. It is thought that most of the feldspar minerals first change to kaolinite ($Al_2Si_2O_5(OH)_4$) from which the removal of silica yields bauxite. An examination of some of the bauxite occurrences of the world will serve to illustrate the geologic conditions for lateritic deposits of this kind.

ARKANSAS BAUXITE DEPOSITS

The principal production area lies about 16 miles southwest of Little Rock, near the town of Bauxite (Fig. 22). Another smaller deposit is only about 1 mile southeast of Little Rock. These districts provide nearly all of the American production, although smaller deposits in Alabama and Georgia have been exploited on a small scale in the past.

The origin of the Arkansas deposits goes back to that period of the geologic past called the Cretaceous. A long cycle of erosion had reduced to a nearly level plain a folded sequence of sediments into which had been intruded two large masses of nepheline syenite. This silica-poor, alumina-rich igneous rock is coarse grained and consists of the minerals *nepheline* ($NaAlSiO_4$), *orthoclase* ($KAlSi_3O_8$), and a minor amount of dark minerals such as *biotite* ($K(Mg,Fe)_3AlSi_3O_{10}(OH)_2$). Nepheline syenite is very susceptible to weathering and even in the unaltered state contains about 19% of aluminum oxide. When the outcrops of these intrusions were eroded to a surface low enough that the products of weathering were not carried off by erosion, the process of lateritization was able to commence. The climate at this time must have been one of high temperature and abundant rainfall. Many geologists believe that the rainfall was seasonal with an annual wet and dry period. In any event, a thick residual deposit of bauxite formed by the action of the weathering solutions, which gradually changed the feldspars and nepheline first into *kaolinite* ($Al_2Si_2O_5(OH)_4$) and then to bauxite (Fig. 23). This was accomplished by the initial solution and removal of sodium and potassium followed by the slower extraction of silica from the kaolinite. Some erosion continually removed the bauxite forming on the higher parts of the region. When the Eocene seas began to encroach on the area, lenses and beds of bauxite gravels were deposited nearby along with other sediments on the sea bottom. This transported bauxite is of neglegible importance compared to the vast amount lying as undisturbed blankets and pockets on the weathered nepheline syenite. As the seas became deeper, the whole area was covered with sedi-

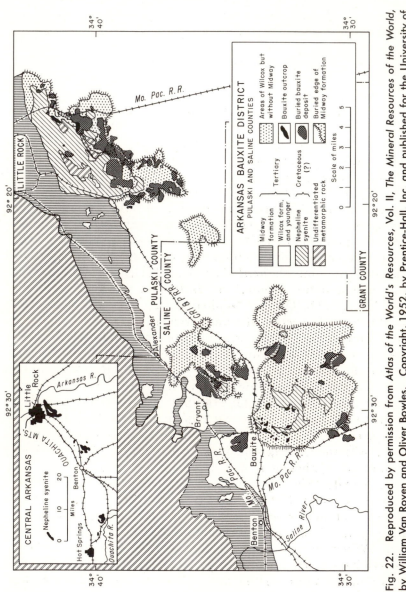

Fig. 22. Reproduced by permission from *Atlas of the World's Resources, Vol. II, The Mineral Resources of the World,* by William Van Royen and Oliver Bowles. Copyright, 1952, by Prentice-Hall, Inc. and published for the University of Maryland. Originally adapted from *U.S. Bur. Mines Rept. of Inves. 4251,* 1948.

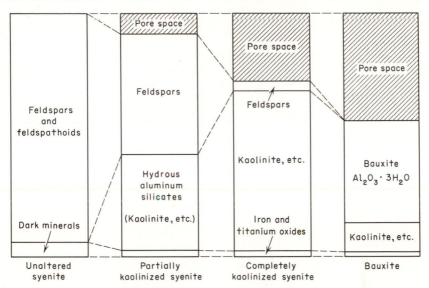

Fig. 23. Diagram showing volume and mineral changes that have taken place when nepheline syenite changed progressively to bauxite ore. Analyses were made from a series of samples taken from a locality near the Lantz mine, Arkansas. (After Mead, *Econ. Geol.*, Vol. 10, p. 48, 1915.)

ments, and the process of lateritization ended. More recent erosion has partially removed this cover of young sediments and exposed the old bauxite deposits to the shovels of the present-day miners (Fig. 24).

The ores are present in an irregular layer up to 35 feet in thickness and averaging 11½ feet. In many places as much as 80 feet of overburden must be removed to uncover the bauxite. The bottoms of the ore bodies are very irregular, and mining must be skillfully done in order not to take the many pinnacles and ridges of kaolin clay and slightly altered nepheline syenite. In many places vertical sections through the ore show the many stages in the progressive alteration of the igneous rock. Starting at the bottom of such an exposure one can see the following substances one on top of another: (1) unaltered nepheline syenite, (2) partly kaolinized syenite, (3) completely kaolinized syenite, (4) bauxite high in silica, and (5) high-grade bauxite ore. In some places, the bauxite gravels in the Eocene sediments can be worked where they are of sufficient purity. The total reserves in this district have been estimated between 50 and 75 million tons.

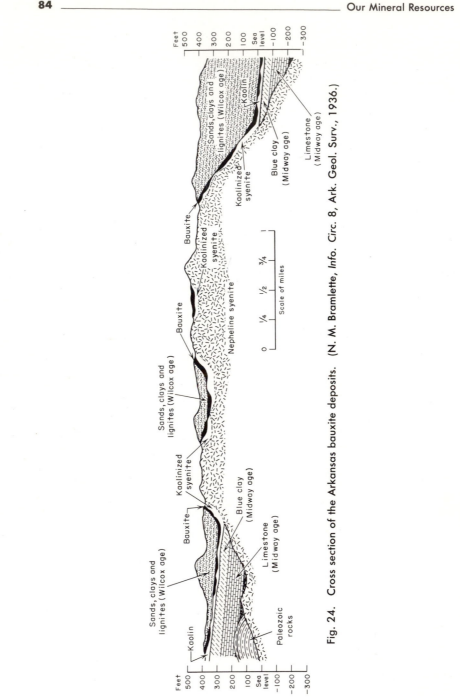

Fig. 24. Cross section of the Arkansas bauxite deposits. (N. M. Bramlette, Info. Circ. 8, Ark. Geol. Surv., 1936.)

BRITISH AND DUTCH GUIANA BAUXITES

These deposits are, at the present time, the major source of bauxite in the world, accounting for over 33% of the recent total production. The ore lies in the dense jungles about midway between the coast and the foothills of the mountains. The bauxite deposits occur in flat-lying beds and lenses up to 40 feet thick and are often covered by as much as 90 feet of overburden. Since the bauxite is more resistant to weathering than the surrounding rock, the deposits commonly stand in slight relief. For this reason, a favorite method of prospecting is to climb a high tree and look for "swells" in the tree-top level. These often mark hills of productive ore.

The bauxite is generally soft and porous and in most deposits highly saturated with water. Large drying plants utilizing rotary kilns have been set up to remove this water before shipment is made. The aluminum oxide content averages about 60%, which is slightly higher than in the Arkansas ores. An even greater advantage is the low content of silica (2.0 to 2.5%). Total ore reserves are not known but are undoubtedly very extensive.

The protore for the lateritic ores has been difficult to determine for the depth of weathering is so great that fresh rock is never seen at the surface. However, inherited textures in the bauxite, slightly weathered mineral grains and a few deep drill holes indicate that schists, gneisses, and many forms of sedimentary and igneous rocks contributed equally to the formation of the bauxite. Although some deposits are probably still in the process of formation, it seems evident that most of the ores formed before the end of Tertiary time, as evidenced by the extensive cover of Late Tertiary sands.

JAMAICAN BAUXITE DEPOSITS

The presence of bauxite in Jamaica was discovered quite by accident in 1942. Sir Alfred H. D'Costa of Kingston was concerned about the fertility of the red soils on part of his 4000-acre cattle ranch near Claremont, so he sent some soil samples to a laboratory in London for chemical analyses. The chemist reported an alumina content of about 50% with only 3% silica, and the soil was recognized as a good quality aluminum ore. By early in 1944 the four North American producers of aluminum had conducted explorations throughout Jamaica and the nearby countries of Haiti and the

Dominican Republic. Their efforts met with amazing success. Over 350,000,000 tons of high-grade ore were discovered, about 90% of which is in Jamaica. To appreciate the importance of this ore, consider the present United States' reserves of about 50,000,000 tons and a rate of consumption of bauxite by United States and Canadian producers of about 4,500,000 tons yearly. The newly discovered ores are of better quality than the remaining United States' reserves.

The Jamaican deposits lie in three separate districts that occur along a northeast-southwest diagonal through the center of the island (see Fig. 25). Their elevations range between 300 and 5000 feet above sea level, and none of the districts is more than 15 miles from the coast, where the water is deep enough that loading docks can easily be constructed. Aerial tramways have been devised for transporting ore to the coast from the St. Ann district.

The ores lie in basin or troughlike depressions in the underlying limestone. Many ore bodies are continuous for several miles, but most are smaller, and the average size is about 10 acres. All of the ore is at the surface so that the only overburden that must be removed is a few inches of sod. Thicknesses of bauxite up to 114 feet have been recorded, but the average is about 25 feet. Each of the three districts is a large basin or valley which has formed as a result of major block faulting (see cross section, Fig. 26). The drainage of these basins is nearly all underground through the porous laterite and into the cavernous limestone below.

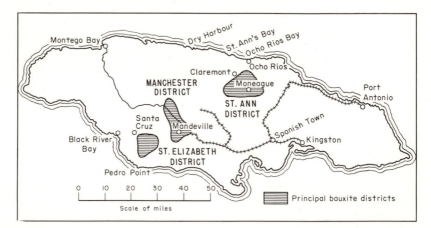

Fig. 25. Map of Jamaica showing locations of three bauxite mining districts. (From Schmedeman, *Eng. and Min. Journ.*, Vol. 151, No. 11, p. 98, 1950.)

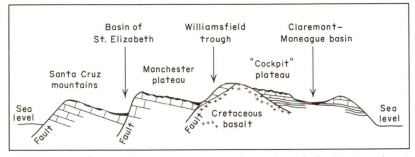

Fig. 26. A cross section from southwest to northeast through the Jamaican bauxite districts (section is 80 miles long). (From Schmedeman, *Eng. and Min. Journ.,* Vol. 149, No. 6, p. 80, 1948.)

The bauxite is highly colored, and shades of yellow, brown, red, and magenta are common. It is very porous and permeable yet exceedingly fine grained (particles range between 0.1 and 1.0 microns in diameter). Drilling and blasting will probably not be necessary for the mining of the bauxite deposits as most of the ore is nearly unconsolidated. The average alumina content of the bauxite is 50%, which is not considered particularly high; but, with an average of only 2.0% silica, this is a sufficient tenor. Iron oxide, which is easily separated from the alumina, makes up about 20% of the bauxite. Nearly 23% of the ore is free water that must be removed in drying ovens before the bauxite is shipped.

The deposits are believed to have formed as residual laterite soils developed as a result of intense chemical weathering of the underlying White limestone, a formation of Middle Tertiary age. It is surprising that this rock has a very low alumina content measured as traces and up to 0.6%. With so small a percentage of alumina as much as 800 feet of the limestone must have dissolved away to leave the amount of bauxite that is present. This figure is not inconsistent when one considers the solution rate of limestone in a humid tropical climate and the length of time since Middle Miocene when solution probably started.

SOUTHERN EUROPEAN BAUXITE

All of the bauxite in this part of the world is the result of lateritic weathering of limestones. The limestones contained impurities of clay which remained as a residual concentration after the calcite dissolved. Continued weathering removed silica from the clay, which gradually changed to bauxite. Of course, limestones do not dissolve along even surfaces but develop caves, sink holes, and other features of ground-water solution. Consequently, the bauxite deposits occupy irregular pockets, lenses, channels, and even cave fillings in the limestones from which they formed.

France, the largest European producer of bauxite, has excellent deposits east of the Rhone valley near the Italian and Swiss borders. The bauxite occurs as irregular beds and pockets on Jurassic and Cretaceous limestones. A cover of younger Cretaceous sediments dates very closely the geologic period when these deposits were exposed to the hot humid climate that changed the residual clay to bauxite. This climate, so different from that of the present, is much easier to imagine when one remembers that at this time the great mountain chains had not yet formed across southern Europe and that low-level lands bordered on warm shallow seas. Reserves of ore with less than 8% silica have been estimated at about 16 million tons.

Hungary, Yugoslavia, Italy, and Greece have deposits of bauxite very similar to those in France, although some of these did not form at exactly the same time. On most deposits a cover of young sedimentary rocks enables the geologist to accurately date their age of formation. In many of the areas, such as in Yugoslavia, the bauxite deposits have been tilted during the mountain-building epochs of the early Tertiary (Fig. 27) and are now exposed in steeply dipping belts, which must be mined by underground methods. Hun-

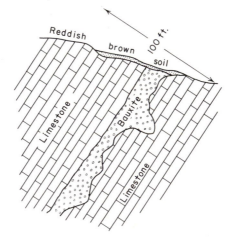

Fig. 27. Cross section of typical bauxite deposit in Dalmatia. (By permission from _The Aluminum Industry,_ by Edwards, Frary, and Jeffries, Chapter by E. C. Harder, Copyright 1930, McGraw-Hill Book Co., Inc.)

gary has the greatest reserves, about 380 million tons, but much of this is lower grade ore. Yugoslavia, Greece, and Italy have nearly 150 million tons combined.

SELECTED REFERENCES

Bramlette, M. N. Geology of the Arkansas bauxite region, *Arkansas Geol. Surv. Inf. Circ.* 8, pp. 1–69, 1936.

Edwards, J. D., F. C. Frary, and Z. Jeffries. *Aluminum and its Production,* McGraw-Hill Book Co., New York, 1930.

Fischer, Elizabeth C. Anotated bibliography of the bauxite deposits of the world, *U. S. Geol. Surv. Bull.* 999, 1955.

Franotovic, Damir. Dalmatia leads Yugoslavia's growing bauxite industry, *Eng. and Min. Journ.,* Vol. 156, No. 12, pp. 78–83, 1955.

Harder, E. C. Bauxite in British Guiana, *Can. Inst. Min. and Met. Bull.,* November, 1936.

Harder, E. C. Stratigraphy and origin of bauxite deposits, *Geol. Soc. Am. Bull.,* Vol. 60, pp. 887–908, 1949.

Hartman, J. A. Origin of heavy minerals in Jamaican bauxite, *Econ. Geol.,* Vol. 50, pp. 738–747, 1955.

Litchfield, L., Jr. The bauxite industry of northern South America, *Eng. and Min. Journ.,* Vol. 128, pp. 243–248, pp. 461–464, 1929.

Mead, W. J. Occurrence and origin of the bauxite deposits of Arkansas, *Econ. Geol.,* Vol. 10, pp. 28–54, 1915.

Schmedeman, O. C. Caribbean aluminum ores, *Eng. and Min. Journ.,* Vol. 149, No. 6, pp. 78–82, 1948.

Schmedeman, O. C. First Caribbean bauxite development, *Eng. and Min. Journ.,* Vol. 151, No. 11, pp. 98–100, 1950.

Singewald, Q. D. Hungarian bauxite, *Econ. Geol.,* Vol. 33, pp. 730–736, 1938.

Zuns, V. A. Bauxite resources of Jamaica and their development, *Col. Geol. and Min. Res.,* Vol. 3, No. 4, pp. 307–333, 1953.

copper

INTRODUCTION

Uses for Copper

Only a few examples must suffice here to illustrate the diversity of the thousands of uses for this metal. The major electrical application of copper is apparent in the extensive webs of power and telephone lines spun by man across the land and even under the sea. Switches, electric-light sockets, and a thousand small parts for many electric appliances are fabricated from copper or its most useful alloy brass.[1] Builders use copper for pipes, gutters, roofing, and hardware. Cooking utensils, floor lamps, ornamental containers, coinage, plates for etching and other printing processes, and military armament are uses which consume great quantities of copper

---

[1] Brass is an alloy of copper and zinc, with the most common proportion being two parts of copper to one part of zinc. Other metals may be added in small amounts.

and its alloys. Many chemicals of copper are used for various industrial processes, for insecticides, and even for medicines. Each of us uses copper directly or indirectly hundreds of times each day in our homes, schools, and offices.

Production of Copper

The United States has for many years been the major producer of copper ore in the world as well as the principal consumer. In 1956 the United States' production of copper was 998,570 short tons, or a little less than 30% of the world output of 3,405,000 tons. In this year Chile was the second most important producer with over 478,000 tons, followed in order by Northern Rhodesia (395,000 tons), Russia (385,000 tons), and Canada (326,600 tons). North America is very fortunate in being so well endowed with this strategic metal.

GEOLOGY OF COPPER

Copper is a metal which commonly occurs in nature with many other metals in nearly every type of ore deposit described in the first chapter of this book. For this reason, almost every country that has any metal production has some recoverable copper. Because of America's leading role in the production of copper, most of the ore deposits which are used as examples for the different geologic occurrences will be domestic ones. Space permits the inclusion of only a few important districts, but those that have been chosen will serve to impress the reader with the great variety of geologic settings in which copper ores are found.

Copper Ore Minerals

There are over 150 different minerals which contain some copper, but of these only eight are of any great importance as ore minerals. Four are classed as sulfides and are primary minerals deposited from hot solutions rising from below. The copper carbonates and oxide minerals are all secondary and result from the weathering of primary copper minerals. Native copper (Fig. 28) probably fits in both categories. The following list names these minerals and describes their chemical compositions:

one inch

Fig. 28. Native copper crystals with calcite.

Mineral Name	Formula	Copper Percentage
chalcopyrite	$CuFeS_2$	34.5
bornite	Cu_5FeS_4	63.3
enargite	Cu_3AsS_4	48.3
chalcocite	Cu_2S	79.8
malachite	$Cu_2CO_3(OH)_2$	57.5
azurite	$Cu_3(CO_3)_2(OH)_2$	55.3
cuprite	Cu_2O	88.8
native copper	Cu	100.0

Native Copper District of Michigan

These deposits are located on the Keweenaw Peninsula, a finger of land that juts into Lake Superior from its southern shore. The presence of native copper was known to the early Indians of this part of the country. Many arrow and spear points, ornaments, and tools were made by these people out of the soft metal. The Jesuit priest-explorers first recorded the presence of copper ore in the seventeenth century, but modern mining did not start until 1845. The district soon became the principal copper-mining area of the North American continent and continued as a major producer for over 80 years. After having yielded over 8½ billion pounds of copper, the old deposits of the district are about mined out, but a new mass of low-grade sulfide ore in Ontonagon County is now being exploited on a very large scale. This is the great White Pine deposit.

The native-copper deposits (Fig. 29) lie in the south limb of the great Lake Superior syncline (see page 66) in a sequence of ancient basalt lava flows and conglomerates 25,000 feet thick. These rocks dip towards the northwest at an angle of about 40°. The same rocks reappear in Ontario along the north shore of the lake, but here they contain no copper. The eastern edge of the district is cut off by a large fault, which brings these rocks into contact with nearly flat-lying younger sandstones. The ores occur as native-copper fillings in gas cavities in the porous tops of many of the lava flows, as cementing material for the pebbles in some of the conglomerates, and in veins which have filled some minor faults in the area. In many of the deposits the ore is remarkably persistent in its extension and consistent in copper content. The Calumet conglomerate has been mined over a length of 18,000 feet and to a depth of 9000 feet down the dip of the bed.

The history of these deposits, as we might reconstruct it, started in the distant geologic past when great flows of basaltic lava spilled out onto the surface of the land. As each lava flow cooled, gases in it formed bubbles which rose to the top. Because the lava cooled more rapidly at the top, these rising bubbles encountered a cooler more viscous layer or even a solid crust and were not able to rise to the surface and burst. Hence, the tops of the flows became very porous and spongy. With many such flows piled one on top of another, there were incorporated into the rock sequence many zones which had high porosity and permeability. During the long intervals of time between some eruptions, streams developed on the surfaces of the older flows and deposited layers of gravels also with high permeability. These beds and flows were later involved in the folding which produced the great syncline. They were tilted to their present attitude, and a major fault truncated their southeastern extent. Sometime after the folding and faulting heated copper-bearing solutions utilized the available rock openings in

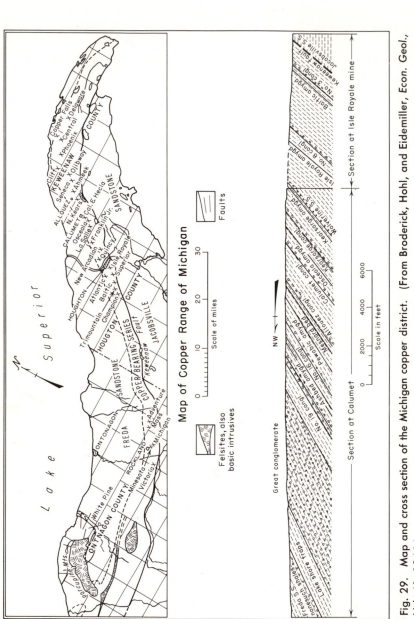

Map of Copper Range of Michigan

Felsites, also basic intrusives

Faults

Scale of miles

0   10   20   30

Section at Calumet

Great conglomerate

NW

Section at Isle Royale mine

Scale in feet

0   2000   4000   6000

Fig. 29. Map and cross section of the Michigan copper district. (From Broderick, Hohl, and Eidemiller, *Econ. Geol.*, Vol. 41, 1946.)

their ascent to the surface. As they traveled through the porous tops of the old lava flows and through the interbedded conglomerates, they deposited many minerals including copper in the rock openings. Mineral fillings of this kind in the round bubble holes of basalts are known as *amygdules,* and the copper deposits formed in this way are called *amygdaloidal lodes. Conglomerate* and *fissure-vein lodes* are the two other types of deposits recognized in the area, and these undoubtedly formed at the same time and in much the same way. Extensive erosion and glacial activity have exposed the deposits and left the surface of the area in its present state. Glacial debris in Wisconsin, Minnesota, Michigan, Iowa, Illinois, and Nebraska contain nuggets of copper plowed up and scattered by the advance of the great ice sheets over these copper deposits.

An interesting feature of the fissure lodes are the large masses of copper that were encountered in them, some weighing as much as 500 tons. Such pieces caused great trouble for the miners, for they would not respond to blasting and had to be literally sawed to pieces. Many beautiful specimens of copper crystals were found in these veins.

One of the great puzzles is why in this particular district the solutions deposited native copper, whereas in most deposits copper occurs combined as sulfides. A widely accepted theory suggests that there may have been considerable sulfur and arsenic carried with the copper in the hot rising solutions, but in the hematite-rich tops of the lava flows the sulfur and arsenic combined with some of the oxygen of the hematite and escaped as gases. This action bleached the red basalt and liberated the copper in the metallic state. This theory is strengthened by other bleached ores of native copper in beds elsewhere stained red by hematite.

A recently proposed and provocative theory[2] suggests that copper was present in the lava flows and crystallized as native copper because the sulfur first escaped as a gas before it could combine with the copper. From intrusive magmas copper and sulfur escape together and combine as sulfides of copper. The theory further suggests that sea water which was incorporated in the sediments and lava flows of the Michigan district was heated and then squeezed out during the crushing and compaction of the beds.

---

[2] Cornwall, H. R., A summary of ideas on the origin of native copper deposits, *Econ. Geol.,* Vol. 51, pp. 615–631, 1956.

These chloride solutions easily dissolved the traces of native copper from the basalt and redeposited it as larger masses in the presence of calcite and zeolites in the amygdaloidal openings and other permeability zones.

Disseminated Copper Deposits

Because they produce more copper than any other type, special consideration should be given to the nature and origin of the disseminated or so-called *porphyry-copper deposits.* The term *porphyry* refers to the texture of the intrusive igneous rocks in which these deposits have developed. (Igneous rocks that are generally fine grained but contain a scattering of coarser crystals are called *porphyries.*)

These deposits are characterized by the following properties: (1) The ore is generally low grade, averaging between 1 and 2% copper. (2) There is commonly a great quantity of ore so that the amount of minable copper is often very considerable. (3) The ore minerals occur as tiny grains scattered along a dense network of tiny fractures in a highly altered igneous rock. (4) The weathering and hydrothermal alteration of the igneous rock usually result in ore which is soft and easy to mine. (5) A zone of enrichment is often present near the surface where secondary chalcocite has been deposited from downward-moving ground-water solutions.

Emmons[3] has suggested a plausible theory for the origin of such disseminated deposits in highly shattered igneous rocks (see Fig. 30). He proposed a sequence of events which may be summarized as follows: (1) A stock is intruded at moderate depths. It is probably a pinnacle of a large batholith at greater depth. (2) Cooling forms a thick upper crust or *hood* of solidified igneous rock at the top of the stock while the interior is still mushy. (3) Magmatic and hydrothermal fluids, not only from the magma of the stock but also from the batholith below, collect under the hood. (4) The internal pressure from these gases finally shatters the hood and the overlying country rock, and the escape of the gases and fluids through the newly formed openings leaves the deposits of ore minerals. (5) Erosion later uncovers the stock, and the action of ground water may cause a zone of supergene enrichment in the original disseminated deposit.

---

[3] Emmons, W. H., *The Principles of Economic Geology,* 2nd ed., McGraw-Hill Book Co., New York, 1940, pp. 286–287.

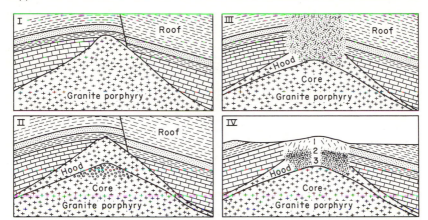

Fig. 30. I. Emplacement of a small stock of granite porphyry, probably part of a larger underlying granitic mass. II. Solidification of the hood—about a mile thick or more. III. Fracturing of hood and roof of intrusion; deposition of low grade hypogene ore or protore. IV. Erosion and enrichment forming (1) a leached zone and (2) a zone of supergene chalcocite ore above (3) the hypogene protore. (By permission from *Principles of Economic Geology*, by W. H. Emmons. Copyright 1940, McGraw-Hill Book Co., Inc.)

BINGHAM DISTRICT, UTAH

This is the largest and most productive copper mine (Fig. 31) in the country with a production of approximately 230,000 tons of copper in 1957, over 6,000,000 ounces of silver, 415,500 ounces of gold, and considerable quantities of molybdenum, lead, and zinc. The district ranks second in the country in gold and molybdenum production and third in the production of silver.

The Bingham mines are situated on the eastern flanks of the Oquirrh Range about 20 miles southwest of Salt Lake City. The main disseminated-copper ore body occupies part of an intrusive mass of porphyry. The magma was injected near the center of a syncline formed in a 10,000 foot sequence of quartzite containing several layers of limestone. The ore solutions penetrated and deposited their load in a shattered zone of the porphyry and created an oval-shaped mass of ore 5000 feet long, 3000 feet wide, and extending to a depth of 2000 feet. Several contact-metamorphic and hydrothermal deposits of relatively small size and value were formed in the beds of limestone which slope up towards the surface from their point of contact with the porphyry intrusion. The area, particularly the limestone deposits, has been complicated by con-

Fig. 31. World famous copper mine at Bingham Canyon, Utah, operated by the Kennecott Copper Corp. This view taken towards the south shows most of the 933 acres being mined. From the bottom to the top level on the west side is a difference in elevation of 2060 feet. One hundred and seventy miles of standard gauge track wind into the mine, and most of this is moved continually to make room for mining. Note many full-size trains on various levels. (Photo courtesy of Kennecott Copper Corp.) (See also the enlarged photograph at the beginning of Section one.)

siderable faulting, but the main disseminated ore body is simple enough to be mined as a terraced open pit into which ore trains enter from tunnels along the side. The ore cars are filled by huge electric shovels that load the rock blasted from the terraces.

The ore (Fig. 32) is a soft gray rock with a "salt and pepper" appearance caused by the very tiny specks of dark ore minerals in the light-colored, highly fractured, and altered porphyry. The main ore minerals are chalcocite and chalcopyrite, with lesser amounts of bornite, molybdenite (molybdenum sulfide), and other sulfides of copper. A large amount of pyrite, which contains most of the gold in the ore, is also present. There is a weathered and bleached zone near the surface, under which is the darker, chalcocite-rich zone of supergene enrichment. The average run of ore contains slightly over 1% copper, but rock with as little as 0.6% copper may be mined profitably. The ore averages about $0.40 per ton in gold and silver, with additional values in other minor metals. It is said that the value of the by-products pays for the mining and processing of the ore and the copper content represents the profit.

Vein-Lode Copper Deposits

This is a very common type of ore occurrence, not only for copper but for other metals as well. Copper is a common by-product from veins worked primarily for gold, silver, lead, and zinc, and these metals are frequent accessories in veins mined for their copper content. The fluids from which such deposits originated were probably identical to those which formed disseminated deposits. However, instead of permeating a dense network of minute fractures, the fluids traveled through and deposited their loads in discreet sheetlike fractures such as fault zones. Under the simplest conditions this results in a tabular ore body, which like a page of this book is large in two dimensions in comparison to its thickness. There are several excellent examples of vein-lode deposits which will be discussed in later chapters, but the great deposit at Butte, Montana, is the principal copper-mining district of this type.

BUTTE DISTRICT, MONTANA

Although the Butte district ranks third in present production, it still might be considered the greatest deposit of copper in the world and one of the outstanding examples of vein lodes. Since modern mining was first started in 1868, it has yielded metals valued at more than two and one half billion dollars. This production includes more copper than has been recovered from any other single district, as well as great quantities of zinc, silver, gold, lead,

Fig. 32. Typical porphyry ore from Bingham Canyon, Utah. Dark specks are ore minerals. Note offset veins.

and manganese.   All of the production has come from a region scarcely more than 2 by 4 miles in area, beneath which are now more than 1000 miles of underground workings.   The district started out as a gold- and silver-mining camp, but deeper workings encountered excellent copper ore, so that since 1892 copper has brought the greatest dollar values.

The mining district lies on the west side of the so-called Boulder batholith, a huge intrusion about 64 miles long and 12 to 16 miles wide at its outcrop.   Near the edge but within this mass of igneous rock three different sets of intersecting faults were formed, probably at about the same time that ore-bearing solutions were rising from the still liquid depths of the batholith.   The solutions altered the igneous rock and deposited ore minerals in the fault fissures (Fig. 33).   At least four sets of later faults contain no ore, but they cut up the ore bodies into blocks which have made the mining problems very complex.   The great Continental fault, the youngest of these, borders the district on the east and has lowered it some 1500 feet.   If it were not for this fault, the whole district might have been eroded away during the comparatively recent periods of intense erosion.   Even so, many gold placers, which led to the discovery of the district, were formed by the erosion debris of great amounts of ore.

Chalcocite, bornite, and enargite are the chief copper-bearing minerals in the ore.   It also contains many rare minerals, some of which were first described from this district.   The ore is very high grade, bearing 4 to 5% copper, and each ton contains 2 or 3 ounces of silver.

Replacement Lodes

Deposits of this kind generally indicate a great depth of formation and rather high-temperature conditions.   In many respects they might be closely compared with contact-metamorphic deposits, especially in their mineralogy and their similar affinity for limestones.   However, for some reason the metamorphic changes and the deposition of ore minerals from the magmatic solutions occurred at a distance from the contact zone.

NORANDA DISTRICT

This great mining district in Eastern Quebec is a good example of a sulfide replacement ore body.   Here thick, Precambrian lava

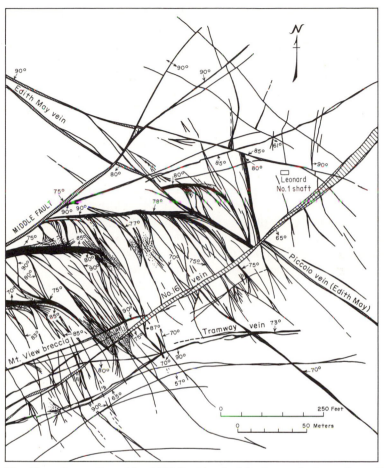

Fig. 33. Plan of a portion of the 1200 foot level of the Leonard mine at Butte, Montana. This shows the complex faulting and so-called horsetail developments. Area is all granite. (From Sales, *A.I.M.E. Tr.*, Vol. 46, 1914.)

flows and pyroclastic beds have been intensely folded into eastward-plunging anticlines and synclines. The rocks have been intruded by thick diabase dikes and by granitic stocks. Nearly vertical faults cut the area. The ores are closely related to the faults and also seem to have been localized by some denser lava flows in the rock sequence. In the ore bodies the volcanic rocks have been completely replaced by sulfide minerals, which are mostly pyrite, pyrrhotite, and chalcopyrite.

The great H ore body is an irregular pipe that extends over 2500 feet below the surface and in cross section is roughly elliptical.

Several small ore masses lie near the H body. The ore contains about 5% copper; and gold, silver, and zinc are important secondary metals.

Oxidized Copper Ores

Such ores as these are always associated with primary copper deposits which have been exposed to weathering. Since the oxidized zone of a copper deposit contains minerals which are generally more soluble than the primary sulfides, oxidized ores only survive in regions which are extremely arid. Even in the most arid climates there is enough moisture to carry downward in solution some of the oxidized copper, so that a zone of supergene enrichment commonly lies below the oxidized ores. Such is the case at Bisbee, Arizona, where much of the ore mined in the early days of the district was from the oxidized zone and consisted mainly of malachite, azurite, and cuprite. Deeper mining exploited the supergene ore and the primary sulfide material. Oxidized ore bodies with zones of supergene enrichment are characteristic of the copper deposits of the Southwest.

CHUQUICAMATA, CHILE

Chile, the second most important copper-producing country, relies on the great deposits at Chuquicamata for over half her annual production. This district contains what is probably the largest known deposit of copper ore in the world, and it is certainly the largest body of oxidized disseminated ore.

The mines are located at an elevation of 9317 feet in the Atacama Desert, about 150 miles northeast of Antofagasta. As in other porphyry copper deposits, a large mass of monzonite porphyry has provided the site of ore mineral deposition. This rock, with the Elena granodiorite, was emplaced along a fault zone between the older Fortuna granodiorite and a sequence of metamorphic sediments and volcanics. Intense shattering has allowed the formation of disseminated ore in an area 2 miles long, 3600 feet wide, and up to 1900 feet deep. The solutions severely altered the igneous rock and deposited primary sulfides and arsenic sulfides of copper. Deep erosion on the flanks of the Andes Mountains rapidly uncovered the primary deposit, and during a moist period of the past much copper was dissolved from the upper parts and formed a rich supergene layer about 600 feet from the present surface. The climate of today is very arid, and air can penetrate deeply into the rock pores. The air caused the oxidation of much of the supergene ore. Consequently, the deposit may be divided from top to bottom into four rather distinct horizons: (1) leached capping, (2) oxide ore, (3) supergene ore, and (4) primary sulfide ore.

At the present time oxide ore is being mined from a large open pit located where the thick leached zone has been removed by erosion. The oxidized material contains over 2% copper in a mineral state that could only exist in such extreme aridity as that present in the Atacama Desert. The ore minerals are a group of very soluble sulfates and chlorides of copper that are easily extracted by solution in water. The sulfide ores, recently exposed in the pit, are now processed in a newly constructed plant that can handle 30,000 tons each day.

The annual production of Chuquicamata is about 270,000 tons of copper, and the estimated ore reserves of over a billion tons assures a long productive life, particularly now that sulfide ores can be used.

Sedimentary Deposits of Copper

It is well known that copper can be deposited during the formation of sedimentary rocks. This soluble metal must certainly reach the oceans in considerable quantities dissolved in dilute solutions from the weathering of copper-bearing rocks and ores. Copper has been detected in small amounts in the bodies and shells of certain marine mollusks, and some microorganisms are believed to be responsible for the deposition of copper sulfides and native copper in marine muds.

KUPFERSCHIEFER DEPOSITS

The famous Kupferschiefer deposits have been the main source of copper in Germany for many centuries, and enormous reserves still remain. The ore horizon is a Permian black shale about three feet thick, which occurs over an area of 22,000 square miles. It is very rich in organic matter and contains up to 3% copper with lesser amounts of other metals. The ore minerals are all sulfides. In the Mansfield district four synclines have an estimated total minable copper content of 3,700,000 tons in an area of 73 square miles.

A hydrothermal origin has been advocated by many geologists

because of the sulfide mineral content and other features. This seems an unlikely hypothesis for such a thin persistent ore bed that underlies so vast an area. A more plausible theory, advocated by most European geologists, is that the metals were deposited with the organic shale in a stagnant, shallow sea.

SELECTED REFERENCES

Broderick, T. M. *Geology, exploration, and mining in the Michigan copper district,* XVI Int. Geol. Cong. Guidebook 27, 1933.

Broderick, T. M., C. D. Hohl, and H. N. Eidemiller. Recent contributions to the geology of the Michigan copper district, *Econ. Geol.,* Vol. 41, pp. 675–725, 1946.

Copper Resources of the World, 2 volumes, XVI Int. Geol. Cong., Washington, 1935.

Cornwall, H. R. A summary of ideas on origin of native copper deposits, *Econ. Geol.,* Vol. 51, No. 7, pp. 615–631, 1956.

Davis, W. *The Story of Copper,* Appleton-Century, New York, 1924.

Deans, T. The Kupferschiefer and the associated lead-zinc mineralization in the Permian of Silesia, Germany and England, *Int. Geol. Cong.* 18th, Great Britain, Rept. Pt. 7, pp. 340–352, 1950.

Hunt, R. N. *Bingham mining district, Utah,* XVI Int. Geol. Cong. Guidebook 17, 1933.

Materials survey, copper, United States Bureau of Mines, Washington, D. C., 1952, 674 pp.

Perry, V. D. Geology of the Chuquicamata orebody, *Min. Eng.,* Vol. 4, No. 12, pp. 1166–1169, 1952.

Sales, Reno H. Genetic relations between granites, porphyries and associated copper deposits, *Min. Eng.,* Vol. 6, No. 5 (*A.I.M.E. Tr.,* Vol. 199), pp. 499–505, 1954.

Sales, R. H., and C. Meyer. Results from preliminary studies of vein formation at Butte, Montana, *Econ. Geol.,* Vol. 44, pp. 465–484, 1949.

Weed, W. H. *Geology and ore deposits of the Butte District, Montana,* U. S. Geol. Surv. Prof. Paper 74, 1912.

White, W., and J. C. Wright. The White Pine Copper Deposit, Ontonagon County, Michigan, *Econ. Geol.,* Vol. 49, No. 7, pp. 675–716, 1954.

chapter 5

lead and zinc

Although these metals differ greatly in their chemical and physical properties, they commonly occur together in nature. With but a few exceptions, ores that are mined for their lead content generally contain commercial quantities of zinc, and the reverse is equally true. For this reason it has been convenient to discuss both of these metals and their geologic occurrences in this one chapter.

Uses for Lead and Zinc

Next to copper, lead and zinc are the most essential nonferrous metals used in modern industry. A few of the more important uses for each of these metals are described below, and the reader can undoubtedly add to the list from the scope of his experiences.

Nearly a third of the lead production is used as one of the lead oxides, red lead, and litharge (yellowish orange), and the lead carbonate called "white lead." White and red lead are used in paints, and litharge is an ingredient of certain kinds of glasses and glazes on pottery. Storage batteries and coverings on electrical cables use nearly a quarter of the supply. Great quantities of lead are burned in gasoline as tetraethyl lead. Plumbers still use some lead pipe, and lead bullets and shot are familiar products. There are many alloys containing lead of which some of the most common are: solder (lead and tin), type metal (lead and antimony), pewter (lead, tin, copper), and Wood's alloy (lead, bismuth, tin, cadmium), which is used in automatic fire extinguishing systems because of its melting point of only 70°C.

Galvanizing, that process in which steel is given a protective coating by dipping it into molten zinc, uses nearly 40% of the zinc produced each year. Zinc castings of automobile carburetors, pumps, and hundreds of other items consume a great tonnage of the metal. Rolled zinc is used for glass-jar tops, battery cans, and similar products. The manufacture of the widely used alloy, brass, requires nearly one-fifth of the zinc marketed each year. Zinc oxide, a common white pigment in paints, and zinc chemicals and medicines are minor uses.

Production

The following table shows the five countries which produce the most lead and also the most zinc. It is remarkable testimony to the universal association of these two metals that the same five countries lead in the production of both metals and nearly in the same order. In the past years Belgium ranked as a major refiner of zinc ores. The figures indicate metric tons of metal. It is gratifying to see that the United States has such abundant resources in these useful and strategic metals.

	Lead 1956	Zinc 1955
United States	352,826	514,500
Australia	333,658	287,300
Mexico	220,029	296,900
Canada	188,971	428,500
U.S.S.R.	290,000*	330,000*
Total world	2,420,000	3,200,000

* Estimated

GEOLOGY

Ore Minerals of Lead and Zinc

Of the many minerals which contain these metals, there are only six that are of great importance as ore minerals. The New Jersey zinc deposits have a different suite of minerals which will be described in a later part of this chapter. Only galena (Fig. 34) and sphalerite (sometimes known as zinc blende) are primary minerals, and all of the others form in the zone of oxidation from the weathering of some deposit containing sphalerite and galena.

Lead Minerals	Chemical Composition
galena	PbS
cerussite	$PbCO_3$
anglesite	$PbSO_4$
Zinc Minerals	
sphalerite	ZnS
smithsonite	$ZnCO_3$
hemimorphite	$Zn_2SiO_3(OH)_2$

Fig. 34. Crystal mass of galena.

Associations

Sphalerite and galena are very common associates, although in some deposits they do occur separately. Galena with silver impurities is called *argentiferous galena*. This is actually one of the most important ore minerals of silver. Lead deposits frequently contain appreciable amounts of bismuth and antimony, and cadmium is a common by-product of zinc ores. Gold and copper minerals are ordinarily present with either lead or zinc ores. Although sphalerite and galena are usually found together, it is common for lead and zinc to "part company" during oxidation when the weathered zinc and lead sulfides pick up oxygen from the air to form sulfates. Zinc sulfate is quite soluble, in comparison to the highly insoluble lead sulfate (anglesite) and is carried away by ground water. The oxidized lead minerals remain as a residue.

Disseminated Deposits

Deposits of lead and zinc that are classed as disseminated form under entirely different circumstances than the copper deposits of the same type. Although both are low grade with the ore minerals sparcely scattered through great quantities of rock, the disseminated deposits of lead and zinc are found in limestones, in contrast to the copper-bearing porphyries. They formed at lower temperatures than the porphyry copper deposits, and replacement of the limestone by ore minerals is a common feature.

TRI-STATE DISTRICT

This large district centers around Joplin, Missouri, but extends into Oklahoma and Kansas (Fig. 35) with a total area of about 2000 square miles. It is the principal zinc-producing area of the country and also yields great quantities of lead. No other metals are recovered from the ore. The low-grade disseminated deposits can only be handled on a paying basis because of the great ore reserves which lie so close to the surface that mining costs are low. Mines rarely are deeper than 300 feet, and many of the deposits can even be mined from open pits. The area has produced nearly one billion dollars worth of metals.

Nearly all the ore from the Tri-State district is recovered from a thick cherty limestone unit called the Boone formation. Before the overlying shales were deposited, the Boone limestones were subjected to a period of weathering that created what is known as

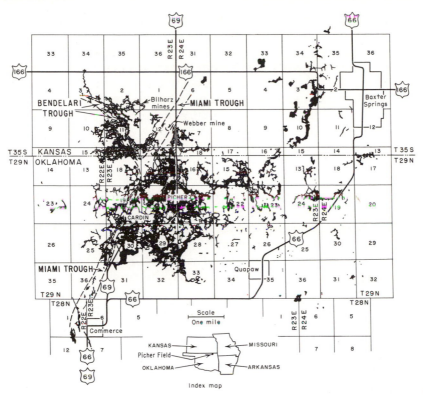

Fig. 35. Map showing the extent of underground workings at the Picher Field, at present the most productive part of the Tri-State zinc and lead district. (From Lyden, *Min. Eng.*, Vol. 187, No. 12, p. 1152, 1950.)

a *karst* topography on the exposed limestone surface. Such a topography is one in which ground-water solutions have developed caves, sink holes, and steep-walled valleys where the roofs of solution tunnels have collapsed. Sink holes and collapse structures were filled with porous, angular limestone and chert rubble. This surface was then covered by younger marine sediments. A period of folding and faulting followed which created very gentle flexures in the form of domes, anticlines, and synclines. Faults with displacements of about 15 to 50 feet are common, and some larger faults have been measured. The significance of these structural features lies in the effects which they had on the very brittle Boone limestones. Tensional stresses attending the folding, and the disruptive forces of the faulting caused large areas of shattering, some of which now contain limestone and chert breccias. The ore solutions naturally followed these permeable zones and others formed by solution activity. They deposited sphalerite and galena in the openings and soaked into the surrounding limestones where the disseminated ore minerals crystallized. The deposits occur as circles around the old sink holes, in elongated bands, and as "sheet" ore confined to certain beds. The mining of these low-grade ores under hundreds of square miles has left much of the area a wasteland of worthless rock and great heaps of mill tailings, called "chat piles" by the local inhabitants.

The origin and nature of the solutions that deposited the ore of the Tri-State district has always been a source of contention among the students of the area. For many years, it was believed that ground water dissolved the metals from distant older sediments and deposited them in the Boone rocks under the conditions described above. This view gradually lost favor and a hydrothermal origin was proposed, even though no adequate igneous source for the ore fluids could be found. Evidence for the hydrothermal theory is not entirely convincing, and the origin of these great deposits is still in doubt. Because nearly all the ore comes from the Boone formation, it has been suggested that the lead and zinc minerals were deposited at the same time as the cherty limestone and later redistributed in some way by solutions that followed permeable trends in faulted and folded structures.

The Tri-State district combined with a lead-producing area in southeastern Missouri, and another lead-zinc district in southwestern Wisconsin and northeastern Illinois called the Upper Mississippi Valley district, make up what is known as the Mississippi Valley lead and zinc metallogenetic province. The geology of the three areas is similar, and the ores probably formed at about the same time.

Massive Replacement Deposits

In many respects, deposits of this kind are similar to copper ore replacement lodes as illustrated by the occurrence at Noranda, Quebec. Such deposits are very high grade where the host rock has been extensively dissolved away and replaced by ore minerals. Of the two deposits used as examples of massive replacement, one has formed in a fine-grained quartzite, and the other in limestone.

SULLIVAN MINE, KIMBERLEY, BRITISH COLUMBIA

This is probably the largest single lead and zinc mine in the world, employing over 1200 miners and mill men. It is located in the Purcell Mountains about 85 miles north of the international boundary with Idaho.

The rocks of the region are Precambrian quartzites and slates which dip steeply to the east along a general north-south trend (Fig. 36). These rocks are probably the same as those which appear in the Coeur d'Alene districts of Idaho, about 250 miles to the south. An intrusion of granite invaded the area about the time when mountain-building forces were crumpling these sediments, which were deeply buried. Mineral solutions from the magma selected the massive bed of quartzite with the highest permeability as a channel for their flow towards the surface. Where the ore-bearing liquids passed through this bed nearly all of the quartzite was replaced by sulfide minerals. Subsequent erosion has exposed the ore body and a small part of the intrusion, which crops out about 5 miles southeast of Kimberley.

The ore, nearly a solid mass of sulfide minerals, consists mostly of iron sulfide (pyrrhotite) but contains about 16% of lead and zinc and 4 ounces of silver per ton. The production is over 6500 tons of ore per day. The replacement character of the ore minerals is strikingly demonstrated by crumpled sulfide layers with sedimentary textures still faithfully preserved by the ore minerals. The ore body is about 6000 feet long where it crops out along the eastern

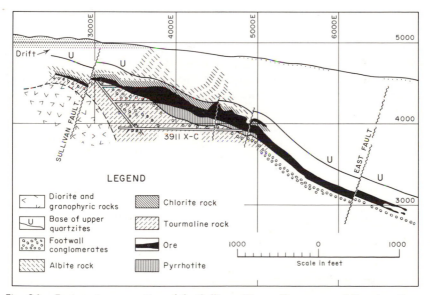

Fig. 36. East-west cross section of the Sullivan Mine. (Swanson and Gunning, *Structural Geology of Canadian Ore Deposits,* Can. Inst. Min. and Met., p. 224, 1948.)

flank of a north-south trending anticline. The maximum width is 270 feet, and the depth to which the ore extends is not known. The abundance of pyrrhotite and other features indicate ore formation at very high temperatures from solutions probably under great pressure.

FRANKLIN DISTRICT, NEW JERSEY

These deposits lie about 50 miles west of New York City in Sussex County, New Jersey. Mining of the zinc ores started over 100 years ago, and the district, though now closed,[1] has been one of the most important producers of zinc in the country. The deposits are highly unique in many respects: (1) The ores contain up to 25% zinc with scarcely any lead or sulfur. (2) Over 100 different species of minerals have been recognized in the area, many of which have been found no place else in the world. Even the principal ore minerals are unusual oxides and silicates of zinc rather than the more common sphalerite. (3) The formation of the zinc ore bodies was the result of special conditions and geologic events seemingly peculiar only to this district.

Reconstruction of the geologic events that led to the present deposits can be done in only a speculative way, and many different versions have been published. One plausible theory states that the first zinc mineralization occurred in Precambrian times when two granitic intrusions caused replacement deposits of zinc sulfides in the Franklin limestone. A period of erosion followed, at which time the original zinc deposits were oxidized and sulfur removed, leaving carbonates and silicates of zinc. Paleozoic sediments were deposited over the tilted and eroded ancient rocks, and sometime after the Ordovician the whole area was intruded again by granites and pegmatites, which further altered the Franklin limestone and changed the zinc minerals to the present suite. It was possibly at this time the small veins of pyrite, sphalerite, and galena were introduced. Erosion has uncovered the ancient beds and a new weathered zone of oxide zinc ore has developed locally at the surface.

In another theory the oxide minerals are said to be primary deposits by Late Precambrian hydrothermal solutions. Certain favor-

---

[1] Although the Franklin mine is exhausted, there is still much ore in the similar Ogdensburg body nearby. It is at present shut down awaiting a more favorable price for zinc.

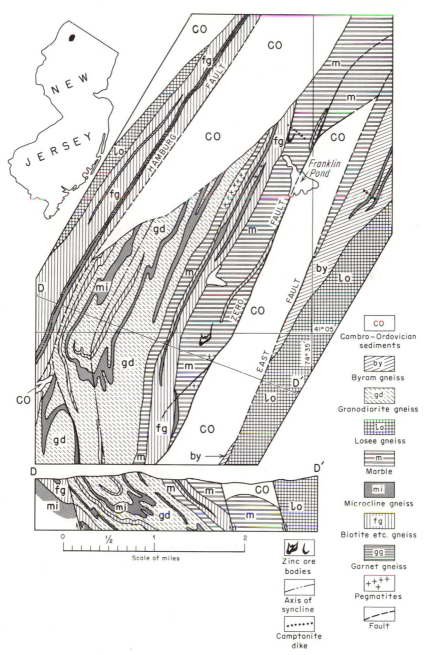

Fig. 37. Geology and structure of the Franklin-Sterling area, New Jersey.   (After Hague et. al., Geol. Soc. Am. Bull., Vol. 67, 1956.)

able horizons were affected in the limestone, which was being folded at the same time as the emplacement of the ore minerals.

There are two ore bodies in the area, both of which lie in synclinal folds of the Franklin limestone (Fig. 37). Their shapes resemble that of a canoe from which one side has been partially removed so that a cross section looks like a hook. The "keels" slope down into the ground at an angle of about 20°, and in the largest body the ore thickness in the keel is as much as 100 feet. The ore itself is very distinctive, containing red zincite (zinc oxide), black franklinite (zinc, iron, manganese oxide), and yellow to brown willemite (zinc silicate), in a matrix of coarse, white crystalline calcite. It is particularly beautiful when viewed in ultraviolet light where the calcite fluoresces with a reddish-orange color, and the willemite a brilliant green.

Most of the franklinite and some of the zincite was used in the manufacture of white zinc oxide pigments for paint. Metallic zinc was recovered from the willemite and most of the zincite. A by-product of the smelting of the ore was an iron alloy called *spiegeleisen,* which contained up to 20% manganese. It was used in the manufacture of steel but is no longer recovered.

Vein and Replacement Lodes

There are many deposits of lead and zinc which like the Butte copper district contain ore that is primarily confined to fissure fillings. Where such veins were deposited from high-temperature solutions a certain amount of soaking into the surrounding rock took place with the formation of additional replacement ores. The rich Coeur d'Alene mining district of Idaho is an excellent example.

COEUR D'ALENE DISTRICT, IDAHO

This area in northern Idaho produces more silver than any other district in the United States. The greatest value in the ore, however, lies in its lead and zinc content. Gold and copper also occur in quantities that warrant their extraction.

The region is one of folded and faulted slates and quartzites of similar age and appearance to those in the district of the Sullivan mine, described above. A small granitic intrusion has been injected into the area, and this was apparently the source of the ore solutions. These liquids filled many of the faults with the valuable ore minerals, which also partially replaced the wall rock. Most of the faults have a west-northwesterly trend, and the ore bodies which they contain are as much as 7000 feet long with an average width of about 9 feet. They have been mined to depths exceeding 5300 feet. Because of the large amount of gouge, the largest faults had permeabilities too low to transmit the magmatic solutions and, consequently, contain no ore. The great Osburn fault has cut the district into two parts, and along this rupture the southern part of the district moved several miles to the west and dropped about 15,000 feet relative to the area north of the fault.

The minerals and replacement character of the ore suggest a formation from high-temperature solutions at great depth. Galena, sphalerite, tetrahedrite, and chalcopyrite are the ore minerals. A variety of gangue minerals, such as pyrrhotite, tourmaline, and magnetite indicate high-temperature solutions. The ore is rich in valuable metals, carrying 3 to 12% lead, 3 to 6% zinc, and 2 to 6 ounces of silver per ton. The formation of oxide ore has been negligible, due perhaps to the abundant moisture of the region and the erosion of the slopes.

Oxidized Ore

Many of the great silver districts of the state of Chihuahua in Mexico are rich in lead and zinc. In this very hot and arid part of Mexico the rich veins, replacements, and stockworks are weathered very deeply. Most of the ore mined in the early days of production from such districts as Santa Eulalia and San Francisco del Oro was oxidized material. Some of the mines were opened by the Spaniards in the middle 1500's and are still major producers today.

SELECTED REFERENCES

Alcock, F. J. *Zinc and Lead Deposits of Canada,* Can. Geol. Surv. Econ. Geol., Ser. No. 8, Ottawa, 1930.

Bastin, E. S., and others, *Lead and zinc deposits of the Mississippi Valley,* Geol. Soc. Amer. Special Paper, No. 24, 156 pp., 1939.

Behre, C. H. Jr., A. V. Heyl, Jr., and E. T. McKnight. Zinc and lead deposits of the Mississippi Valley. *Int. Geol. Cong. Rept. 18 sess.* (1948), Pt. VII, pp. 51–69, 1950.

Hague, J. M., J. L. Baum, L. A. Herrmann, and R. J. Pickering. Geology and structure of the Franklin-Sterling Area, New Jersey, *Geol. Soc. Am. Bull.,* Vol. 67, pp. 435–474, 1956.

Kerr, P. F. *Zinc deposits near Franklin Furnace, New Jersey,* XVI Int. Geol. Cong. Guidebook 8, 1933.

Lyden, J. P. Aspects of structure and mineralization used as guides in the development of the Picher Field, *Min. Eng.,* Vol. 187, No. 12, pp. 1251–1259, 1950.

Pinger, A. W.   Geology of the Franklin-Sterling Area, Sussex Co., New Jersey, *Int. Geol. Cong. Rept. of 18 sess.* (1948), Pt. VII, pp. 77–87, 1950.

Ransome, F. L., and F. C. Calkins.   *The geology and ore deposits of the Coeur d'Alene district, Idaho,* U. S. Geol. Surv. Prof. Paper 62, 1908.

Schofield, S. J.   *Geology of Cranbrook Map-Area, British Columbia,* Can. Geol. Surv. Mem. 76, 1915.

Shenon, P. J.   Lead and zinc deposits of the Coeur d'Alene District, Idaho, *Int. Geol. Cong. Rept. 18 sess.* (1948), Pt. VII, pp. 88–91, 1950.

Stoiber, R. E.   Movement of mineralizing solutions in the Picher Field, Oklahoma-Kansas, *Econ. Geol.,* Vol. 41, pp. 800–812, 1946.

Swanson, C. O., and H. C. Gunning.   Sullivan Mine, *Structural Geology of Canadian Ore deposits—a symposium,* Can. Inst. Min. and Met., pp. 219–230, 1948.

Tarr, S. A.   Southeastern Missouri lead deposits, *Econ. Geol.,* Vol. 31, pp. 712–754, 832–866, 1936.

tin and tungsten

chapter 6

Tin

Uses for Tin

Most of the tin used in this country goes into the manufacture of tin and terne (tin-lead alloy) plate. Other alloys, such as solder (tin, lead), bronze (copper, tin) Babbitt metal (lead, tin, antimony), and type metal (lead, tin, antimony) have hundreds of familiar industrial uses. Tin foil, chemicals, ceramics, miscellaneous alloys, and special plumbing fixtures are all common applications of this useful metal. Another familiar use of tin is in the collapsible metal tubes, which are used as containers for tooth paste, shaving cream, etc. During the tin shortage brought on by World War II a technique was developed whereby such tubes could be made of lead with an inner coating of tin.

117

Production

Southeast Asia is the world's main source for tin. The 1956 production figures show that the countries of Malaya, Indonesia, China, Thailand, and Burma yielded nearly 63% of the world's output of 177,600 long tons. Of these countries, Malaya accounted for 35% of the total with over 62,300 tons of tin. Bolivia is the only major producer in the western hemisphere and was second in the world with a little over 26,000 tons. The Belgium Congo mined ore containing about 14,500 tons of tin. The complete reliance of the United States upon imports was reflected in the severe curtailments in the uses of tin during World War II when the Japanese had overrun Southeast Asia and shipments from South America were uncertain.

Geology of Tin

General relationships. The following characteristics are true of nearly all primary deposits containing tin: (1) The ores are always associated with _granitic_ igneous rocks. (2) The deposits all give evidence of having formed at high temperatures and at great depth. (3) Tungsten minerals are commonly associated with the tin ores. (4) Placers and deposits of surficial enrichment usually mark the surface outcroppings and are often of greater economic importance than the primary ore.

High-temperature veins, contact-metamorphic deposits, and pegmatites are the main types of tin-bearing mineral deposits. In the important districts all of these occurrences are present in varying degrees of importance, and often associated placers or residual concentrations contain much of the marketable ore. For this reason, no attempt has been made here to use examples of the tin districts to illustrate any one type of occurrence.

Ore minerals. _Cassiterite_ (tin oxide) is the main ore mineral. It is a hard, heavy, mineral with a nonmetallic luster and a brown or reddish-brown color. It carries 78.6% tin. The only other important ore mineral of tin is _stannite_ (copper-iron-tin sulfide) with 27.5% of tin. Stannite is a soft grey mineral with a metallic luster and a specific gravity of about 4.4 (that of cassiterite is close to 7.0).

CORNWALL DEPOSITS

Although this district is nearly mined out at the present time, its historic interest warrants a short discussion. The area was one of the principal sources of bronze for early man because of the copper associated with the tin ores. It has been estimated that since 500 B.C. 3.3 million tons of tin have been removed. Most of the early production was from placer deposits, but since 1600 extensive mining of the primary ore has been carried on.

The Cornwall tin and copper deposits occur in and around several granite intrusions. Near the contact zones both the intruded sediments and the granite have been extensively altered. Veins in the sediments and in the granite carry most of the tin, but pegmatites and disseminated contact-metamorphic deposits are also important. The ore solutions escaping from the cooling magma sought out any available openings. At a time when the tops of the intrusions had apparently crystallized to solid rock, the contact zones and cracks in the newly formed granite provided openings that became filled with vein minerals. The surrounding areas were extensively changed by the solutions which even converted granite into a rock consisting principally of quartz and white mica with some topaz and cassiterite. Such rock, called *greisen,* is the most common tin ore of the district. The average ore mined in later years contained about 1¼% tin.

An interesting feature of the area is the zonation in the occurrence of the various metals. This zonation is apparent on the map of the district as irregular bands of deposits containing different metals arranged concentrically around the contact zone. It is even more apparent as one traces the mineral content of some veins from the surface down to great depths and approaches the contact with the igneous rock. In the zone farthest from the contact, where the lowest-temperature deposits formed, the ores contained sulfides of lead and silver with some zinc. Nearer to the contact, or at intermediate depths, sulfides of copper occurred with some tungsten minerals and a little cassiterite. The high-temperature zone nearest the contact and those deposits within the igneous rock itself yielded ores containing cassiterite and some tungsten minerals.

BOLIVIAN TIN DEPOSITS

This great mining district lies in southwestern Bolivia on a plateau nearly 500 miles long by 100 miles wide at elevations between 12,000 to 20,000 feet (Fig. 38). The inaccessibility and poor transportation permits mining of only the highest-grade materials, but there are tremendous reserves in lower-grade deposits which some day will be exploited.

Most of the tin ores of the district are produced from veins

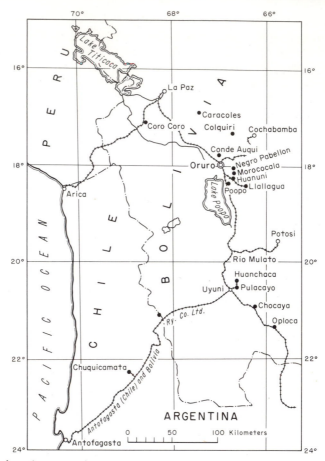

Fig. 38. Location map showing important Bolivian mining areas. (After Chace, *Econ. Geol.*, Vol. 43, 1948.)

formed by high-temperature fluids given off by the many granitic intrusions which probably underlie most of the region. Only small outcrops of the granite appear at the surface from place to place, indicating that erosion has proceeded just far enough to barely uncover the intrusions yet has truncated the highly altered and veined country rock that makes up the "roof" of the intrusion. Because ore fluids travel upwards, the roof rock of an intrusion contains nearly all the ore deposits derived from the magma. Although cassiterite is the principal ore mineral, much of the tin from Bolivian ores is found in the mineral stannite; and a variety of other ore minerals yield copper, silver, bismuth, lead, zinc, tungsten, and molybdenum as by-products. The ores shipped at the present time carry about 2 to 4% of tin. A thick zone of oxidation with a residual enrichment of tin is present, and such ore produced in earlier days commonly contained up to 12% tin. A complex zonation of metal occurrences has also been recognized in this district.

TIN DEPOSITS OF THE MALAY PENINSULA

Extending for more than 1000 miles from near Rangoon southward through eastern Burma, Siam, the Malay Peninsula, and to the islands of Banka and Billiton is the greatest tin-producing area of the world (Fig. 39). In this tropical land of intense erosion and

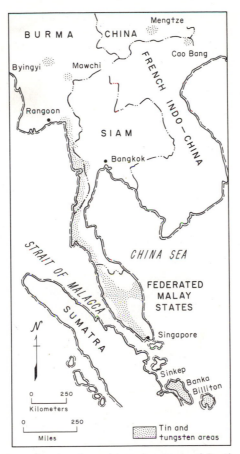

Fig. 39. Location map showing tin and tungsten areas of Southeast Asia. (By permission from *Principles of Economic Geology*, by W. H. Emmons. Copyright 1940, McGraw-Hill Book Co., Inc. Adapted from Jones, 1925.)

abundant rainfall, rich placer tin deposits have formed along nearly every stream on the western side of the mountains, and most of the world's tin production has come from these deposits and from the residual concentrations at the sites of the deeply weathered primary deposits. Many of the secondary deposits are very rich, but over 20% of the tin is mined by huge dredges (Fig. 40) which eat their way through the gravels and profitably recover tin from material too lean to work in any other way. These dredges are like floating factories into which the lean gravels are scooped at one end and put through a series of concentration processes to extract the pure cassiterite. The tailings are dumped out the rear. Where the ground-water table is high, these dredges dig their own channels to float in and continually fill the channel behind themselves.

The primary deposits of tin cluster in and around the series of granite intrusions which extends the length of the province. High-temperature veins, pegmatites, contact-metamorphic deposits in limestones and schists, and greisen alterations in the granites all carry tin. The deep weathering of the primary deposits has resulted in very rich residual concentrations, and placers have formed along streams draining these areas. Cassiterite is hard and heavy and practically unaffected by the chemical and physical decomposition that rapidly deteriorates the rock in which it is enclosed. It is,

Fig. 40. Tin dredge working on the outskirts of Kuala Lumpur, Selangor. It is like a hugh floating factory. Ore is taken in on the left, and the tailings are distributed by the long conveyor on the right after the cassiterite has been removed. (Photo courtesy of Pacific Tin Consolidated Corp.)

therefore, well suited to all natural processes of secondary concentration. Nearly all of the ores contain some tungsten, and the proportion of this metal increases in ores from the northern part of the province where tin is but a by-product to the recovery of tungsten.

Except for the highly efficient dredges, the mining methods would seem to the casual observer rather primitive. In a region where jungles, rain, and mud make heavy machinery difficult to operate and maintain, cheap manpower has supplied the answer. Thousands of coolies carrying ore in baskets over their shoulders do the work of many power shovels and a fleet of trucks. Where the ore is soft enough and a plentiful supply of water is present, the mining is done on steep banks with powerful jets of water. The runoff containing the disintegrated ore is directed into long troughs called *sluices* (Fig. 41). In the bottom of the sluices are strips of wood extending crosswise, behind which the heavy cassiterite particles collect while the lighter particles of rock are washed over the top. This is called *hydraulic mining.* No matter what the method of mining, most concentration of the ore is accomplished in the sluice boxes or *palongs,* as they are locally called. In some modern plants and on the dredge's, vibrating tables called *jigs* receive the flow of water and ore. The motion of the table moves to a collecting trough all the concentrate trapped behind the many ridges on the table surface, while the gangue is washed away.

Tungsten

Uses for Tungsten

We are all familiar with tungsten filaments in electric light bulbs which glow with a white heat without melting. Because of this ability to withstand heat, most tungsten is used in the manufacture of high-speed cutting steels used in drills and other machine tools. Such tools will retain their hardness and cutting edges even at dull-red heat. An alloy of tungsten, cobalt, and chromium, called *stellite,* is widely used for hard facing other metals. Armor plate, guns, and armor-piercing projectiles are all made of tungsten steels. Many electric and electronic devices use tungsten as switch contacts, filaments in electronic tubes, X-ray targets, etc. When

Fig. 41. Sing Foong Mine, Ampang, Selangor, Malaya. The house contains a gravel pump which receives the ore washed down by hydraulic workings and pumps it to the top of the high framework. This is the "palong" or sluice which takes out the tin minerals and from which the tailings are distributed. It is built high to accommodate the heaps of tailings that pile around it. (Photo courtesy of Pacific Tin Consolidated Corp.)

tungsten is combined with carbon, it forms tungsten carbide, one of the hardest substances known other than diamond.[1] Its use in abrasive wheels and cutting tools is steadily increasing.

Production

China is the world's principal source of tungsten ore. In 1956 she produced an estimated 20,000 tons of ore concentrate (60% tungsten oxide) which was a little less than 25% of the total world output of 82,000 tons. The United States, although in second place with 13,600 tons, is still highly dependent upon imports of

---

[1] A newly developed compound, boron nitride, is said to be even harder than diamond.

this metal. Most of our production is from California, Nevada, and North Carolina. Bolivia, Portugal, Korea, and Burma also have significant production.

Geology

General relationships. As with the ores of tin, tungsten deposits occur only where mineralization has taken place at high temperatures and pressures. The ore solutions seem always to have been derived from granitic magmas. It is common for both metals to be present in the same deposit. Pegmatites, high-temperature veins, contact-metamorphic deposits, and secondary concentrations furnish most of the ore.

Ore minerals. There are only two common ore minerals of tungsten: *scheelite* ($CaWO_4$) and *wolframite* ($(Fe, Mn)WO_4$). The former was named in honor of K. W. Scheele, who originally recognized the existence of tungsten. Wolframite was named for *wolframium,* the Latin name for tungsten. Both are exceedingly heavy minerals and at one time were thought to be tin-bearing.

CHINA

The major tungsten-producing area in the world lies in southeastern China in the Nanling Mountains 200 miles northwest of Canton. The deposits were not discovered until 1915, but by 1919 China had become the major producer of tungsten, a position undisputed to the present day.

In the area there are 17 separate mining districts which contain over 80 mining localities. The deposits are veins which have formed in the granite intrusions and in the nearby sediments. Some of the veins are as much as 10 feet wide, over 3000 feet long, and have been mined to depths up to 1200 feet. There are many pegmatites closely associated with the veins. The wall rock of the veins has been intensely altered, and much of the granite has been converted to greisen. The principal ore mineral is wolframite, but some scheelite is present and a little cassiterite. Enough bismuth occurs in the ore to make a valuable by-product. The reserves have been estimated at about 2 million tons of ore concentrate carrying 60% tungsten oxide. Much of the early production was from placers and residual concentrations.

MILL CITY, NEVADA

Since the major tungsten deposits of China have fallen into the hands of the Communists, higher prices have stimulated the exploration for new deposits in this country and the production of lower grade ore from the older districts. There are no great and rich tungsten occurrences in the United States, but many small deposits, mainly in California, Nevada, Colorado, and North Carolina produced an estimated 7.5 million pounds of tungsten concentrates in 1957. One of the older and most productive districts is near Mill City, Nevada.

In this area several granitic intrusions have invaded steeply dipping shaly sediments which contain several thin beds of limestone. Where these beds are inclined towards the surface from the contact with the igneous rocks, solutions from the magma have replaced the limestone beds and changed them into skarns rich in garnet, quartz, and other metamorphic minerals. The scheelite scattered through this material is light brown in color and difficult to recognize in ordinary light. However, the ultraviolet light used by the miners causes it to fluoresce with a bright blue color. The ore averages between 1 and 1½% of tungsten oxide.

The beds are 5 to 6 feet thick and are ore-bearing to 2000 feet from the contact zone. At the surface they extend as outcrops up to 1000 feet in length. Some have been mined to depths of 1400 feet. Although the ore bodies are veinlike in their tabular shape, they are contact-metamorphic deposits in which the solutions have selectively replaced the beds of limestone while leaving the shaly beds largely unaltered. It was fortunate that the solutions were confined to a few thin beds of limestone rather than a thick sequence of limestone, for the same amount of scheelite scattered through a greater volume of rock would not have been rich enough to mine. Much faulting and later intrusions have considerably complicated the geologic picture described above.

PIEDMONT TUNGSTEN DEPOSITS

These deposits in the southeastern states were discovered in the 1940's and now contribute a significant amount of the tungsten produced in the United States.

A number of occurrences of tungsten mineralization have been found in the Piedmont province, from Alabama to Virginia, but North Carolina has most of the productive deposits. The Hamme district in North Carolina lies near the Virginia border. Here quartz veins carry huebnerite ($MnWO_4$) and scheelite ($CaWO_4$) with a variety of gangue minerals. The veins occur in a granodiorite which has been badly sheared near its contact with Paleozoic schists. The veins are lenses up to 400 feet long which lie in the shear zone. Mineralization appears to be the result of high-temperature solutions. Ore bodies average between 0.5 and 1.0% WO_3.

SELECTED REFERENCES

Chace, F. M.   Tin-silver veins of Oruro, Bolivia, *Econ. Geol.,* Vol. 43, pp. 333–383, 435–470, 1948.

Espenshade, G. H.   Occurrences of tungsten minerals in the Southeastern States, *Symposium on Mineral Resources of the Southeastern States,* University of Tennessee Press, pp. 57–66, 1950.

Fitch, F. H.   The tin mines of the Pahang Consolidated Co., Ltd., *Inst. Min. and Met. Tr.,* Vol. 57 (1947–48), pp. 195–247, 1951.

Gibson, R., and F. S. Turneaure.   Tin deposits of the Monserrat mine, Bolivia, *Min. Eng.,* Vol. 187, No. 10, pp. 1071–1078, 1950.

Hosking, K. F. G.   Primary ore deposition in Cornwall, *Roy. Geol. Soc. Cornwall Tr.,* Vol. 18, Pt. 3, pp. 309–356, 1951.

Jones, W. R.   *Tin Fields of the World,* London, 1925.

Kerr, P. F.   *Tungsten mineralization in the United States,* Geol. Soc. Amer. Mem. 15, New York, 1946.

Kerr, P. F.   Geology of the tungsten deposits near Mill City, Nevada, *Univ. Nevada Bull.,* No. 2, Vol. 28, 1934.

Li, K. C., and C. Y. Wang.   *Tungsten,* 2nd ed., Amer. Chem. Soc. Mon. 94, Reinhold Publishing Corp., New York, 1947.

Mantel, C. L.   *Tin,* 2nd ed., Amer. Chem. Soc. Mon. 51, Reinhold Publishing Corp., New York, 1949.

Scrivenor, J. B.   *Geology of Malayan Ore Deposits,* London, 1928.

chapter 7 gold and silver

Gold

Gold is a metal that has played an important role in shaping man's destiny through the ages. It has held a special fascination ever since the first shiny yellow nugget was found by some prehistoric man. Because of its beauty and scarcity, gold has always been a symbol of affluence and power, and its acquistion has often been considered more important than the laws of human decency. Wars of conquest, piracy, murder, slavery, and other acts of human debasement have been committed in furtherance of man's greed for gold. Despite its evil influence, the quest for gold has greatly stimulated exploration and the continued search for mineral resources. In fact, the discovery of America and the colonization of Alaska might have been delayed many decades, if not centuries, had it not been for this stimulus.

Fineness of Gold

The purity, or fineness, of gold is measured in terms of carats. This is an arbitrary scale in which pure gold has a fineness of 24 carats. It is usual to alloy other metals with gold to give a desired shade of color or to impart hardness. Such an alloy is commonly stamped with a designation, such as 12 carat, which enables one to determine the percentage of gold in the alloy used. Twelve carat means $^{12}\!/_{24}$ or 50% pure gold; 18 carat means 75% pure gold. This term should not be confused with the carat used as a measure of weight for gem stones. Another scale of fineness sets the value for pure gold at 1000. Thus 750 fine gold is 75% pure.

Uses

The principal use for gold is currency, although at the present time most of it is stored as bullion to back up in value the paper certificates in circulation. The manufacture of jewelry requires great quantities of gold, most of it alloyed with other metals to increase its strength and hardness. Some of the variety of colors that distinguish these alloys are: "white gold" with silver, platinum, and nickel; "green gold" with silver and cadmium; "red gold" with copper; and "purple gold" with aluminum. Gold leaf has a great number of uses such as lettering on book bindings, sign lettering, and gilding. Some other uses for this metal are gold plating, decorative glaze on china, glass making, chemicals, and dentistry.

Production

In 1957 the gold mines of the world, excluding the U.S.S.R., yielded a total of 30 million troy ounces (12 troy ounces equal 1 pound). The deposits in central and south Africa (Union of South Africa, Ghana, Southern Rhodesia, and the Belgium Congo) produced about 60% of the world's total, with over 17.2 million ounces coming from the Union of South Africa alone. Although the production figure for Russia in 1957 is not known, it is estimated at 9 million ounces. Canada's mines yielded 4.4 million ounces. The combined United States' and Alaskan production of about 1.8 million ounces puts us in fourth place. Australia, Colombia, and the Philippines also produced significant quantities of gold. Most of the United States' yield in 1956 came from the Black Hills

(458,000 ounces) and as a by-product from the copper ores of Bingham, Utah (407,000 ounces).

Geology of Gold Deposits

Ore minerals of gold. Native gold, the most important ore mineral of gold, may occur as submicroscopic specks, thin wire, or larger masses which become "dust" or "nuggets" following the rounding action of stream transportation. Most native gold contains appreciable amounts of silver as an impurity. Many ores have no visible gold; the gold is present only as impurities in such sulfides as pyrite and chalcopyrite. Usually these two minerals contain no gold, but, because of their brassy and golden colors, their presence has often given false hope to the inexperienced prospector. For this reason, they have been called "fool's gold." A scratch of the knife blade quickly brands them for what they are, for pyrite is too hard to scratch and chalcopyrite crumbles to a black powder. Gold is soft and sectile and retains its metallic character even when deeply grooved by the blade.

A few rare minerals contain gold and silver in chemical combination with tellurium. The tellurides *sylvanite* ($AuAgTe_4$) and *calaverite* ($AuTe_2$) are locally important as ore minerals of gold and silver.

Geologic occurrence of gold. Gold is found in commercial amounts in nearly all of the types of ore deposits described in Chapter 1. It is actually a very common metal which can be encountered in minor amounts in almost every geologic circumstance. Much gold is produced as a by-product from ores of other metals. Analyses of sea water show enough dissolved gold to make up an estimated 10 million tons, if it could be extracted from all the oceans. Because gold is so chemically inert, heavy, and resistant to mechanical abrasion, it is admirably suited to all kinds of secondary enrichment except by supergene solutions. Much of the world's production of gold has come from placer deposits (Fig. 42). Only a few examples of the many types of primary and secondary gold deposits are given below.

HOMESTAKE MINE, BLACK HILLS, SOUTH DAKOTA

This great mine, the major gold producer in the western hemisphere, is located at Lead, South Dakota, in the north-central part of the Black Hills. The deposit was discovered in 1876 by a young

Fig. 42. Hydraulic gold mine in placer deposit not far from the Klondike fields. Pipe near horizon carries water from a dam upstream. With enormous pressure water washes gravel from bank and carries it through sluice in foreground where gold is separated.

French prospector named Moses Manuel who later sold the claim for $70,000. Since 1879, when significant production commenced, the deposit has yielded over 500 million dollars worth of gold.

The geology of the deposit can be summarized as a high-temperature hydrothermal replacement of a silicate marble and schist at the crest of a plunging anticline and beneath an impermeable cap rock. The geologic history of the region which culminated in the formation and exposure of this great deposit has been interpreted as follows:

A 9500-foot sequence of ancient sediments and igneous sills was tightly folded into a large complex anticline and syncline during Precambrian times. These folds measure nearly 6000 feet from crest to trough and plunge to the southeast at an angle of about 40°. Extensive erosion truncated the rocks to a nearly level surface which formed the base for an accumulation of Paleozoic and Mesozoic sediments that built up to over 6000 feet in thickness. During the early part of the Tertiary, probably in Eocene times, the whole northern part of the Black Hills suffered numerous intrusions of rhyolite dikes which were injected into the ancient crumpled rocks and probably into the overlying younger sediments. The gold-bearing solutions apparently came from the rhyolite magmas, were funneled upward beneath the inverted trough of the impermeable Ellison beds, and soaked into the underlying Homestake formation (Fig. 43). Finely disseminated gold was deposited along with arsenopyrite (FeAsS), pyrrhotite (Fe_7S_8), and minor amounts of other sulfides. The uplift of the Black Hills was probably contemporaneous with or shortly followed the intrusions of rhyolite. The increased elevation made possible the stripping away by streams of the great thickness of young sediments, exposing the rich ore deposit to the sharp eyes of Mr. Manuel.

The ore is remarkably uniform and has been mined to depths over 5000 feet with no apparent limit to the reserves. When one considers the average tenor is about $16.25 per ton, or a little less than 0.5 ounce of finely divided gold per ton (Fig. 44), it is little wonder that many of the Homestake miners have worked for years without seeing any gold in the ore. The surface plant handles about 4000 tons of ore per day, making a production of a little less than $65,000 worth of gold during each day.

PORCUPINE DISTRICT, ONTARIO

The Porcupine district is the principal gold-producing area of of Canada with 36 mines yielding about 1 million ounces per year. The largest of these is the Hollinger Mine, at one time the world's greatest. It is located north of Sudbury and about midway between

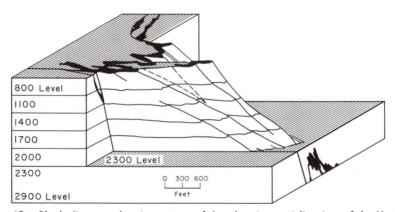

Fig. 43. Block diagram showing nature of the plunging anticlinorium of the Homestake deposit. Ore localized in crests of folds is indicated in black. (From D. H. McLaughlin, _Eng. and Min. Journ.,_ Vol. 132, 1931. Copyright McGraw-Hill Book Co.)

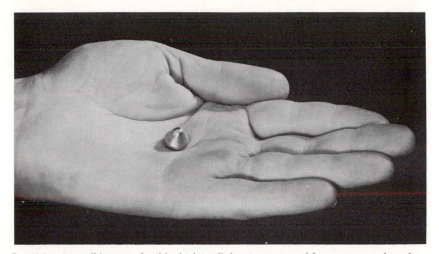

Fig. 44.   A small button of gold which is all that is recovered from one ton of ore from the Homestake Gold Mine.   (Photo courtesy Homestake Mining Company.)

the southern tip of James Bay and Lake Huron (Fig. 45).   Like most of the important mining areas of Canada, the Porcupine District lies in that vast area of ancient and complex crystalline rocks called the Precambrian Shield, which makes up nearly all of the eastern half of Canada and extends into Minnesota, Michigan, Wisconsin, and New York.

The geology of the district is exceedingly complex, for the ancient rocks have been twice folded, intruded by igneous masses, faulted, and eroded (Fig. 46).   The oldest rocks in the region are a series of lava flows which were tilted, eroded, and covered by younger sediments.   All of these rocks were then intricately folded into the complex Porcupine syncline and later intruded by granitic magma which formed pipelike igneous masses now exposed in areas as large as 5000 by 1500 feet.   Fracturing of the country rock, attending or perhaps soon after the intrusions, provided openings for the flow of high-temperature hydrothermal solutions.   These fissures became filled with milky vein quartz that contains the gold and other ore minerals.   The replaced and altered wall rock surrounding many veins can also be mined for its gold content.   Later, basic dikes cut all of the older rocks and ore bodies.   Extensive erosion through much of subsequent geologic time has exposed the ores and destroyed the upper parts of the deposits.

The most important ores of the area come from quartz veins and lenses which are about 10 feet in thickness and as much as 2000

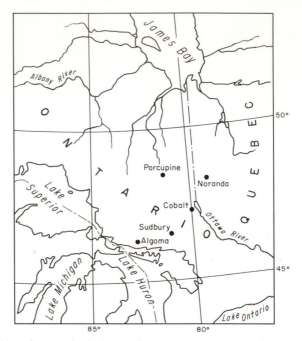

Fig. 45. Map showing the location of some important mining districts in Ontario.

feet in length. There are also some irregular replacement lodes and pipelike ore bodies. The quartz ores carry considerable pyrite, siderite, tourmaline, and sericite with the gold. Minor amounts of other ore and gangue minerals are also present. The gold, usually associated with the pyrite, is commonly in sizable masses. For this reason special precautions must be taken to prevent "high grading," a practice whereby exceedingly rich pieces of ore leave the mine secreted on the persons of the more observant miners.

GOLD DEPOSITS OF THE WITWATERSRAND SYSTEM, TRANSVAAL, UNION OF SOUTH AFRICA

Gold was first discovered in rocks of the Witwatersrand System in 1885, and up to 1957 the many mines of the Rand district have produced an estimated 10.5 billion dollars worth of gold. This is by far the most important gold-mining region in the world, for at the present time about one-third of the world's annual production is from the Rand. Except for a few short breaks, there is a curved belt 90 miles long of gold-bearing rocks in which about 40 mines are active (Fig. 47).   These are some of the deepest mines in the world, many operating at depths between 3000 and 6000 feet, and some will be worked down to 9000- or 10,000-foot levels.

The Witwatersrand is a sequence of Pre-cambrian sedimentary rocks that lies on the eroded surface developed upon ancient gran-ites and schists.   The sequence is divided into two units which are separated by an unconformity.   The Lower Witwatersrand is about 15,000 feet thick in Central Rand and comprises shales, quartzites, iron formations, and a few conglomerates but contains only a little gold.   The Upper Witwatersrand consists of about 9000 feet of sedi-mentary rocks with more quartzites and less shales than in the

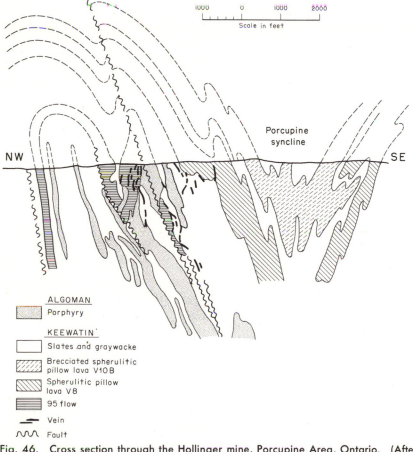

Fig. 46.   Cross section through the Hollinger mine, Porcupine Area, Ontario.   (After Jones, *Structural Geology of Canadian Ore Deposits,* Can. Inst. Min. and Met., p. 467, 1948.)

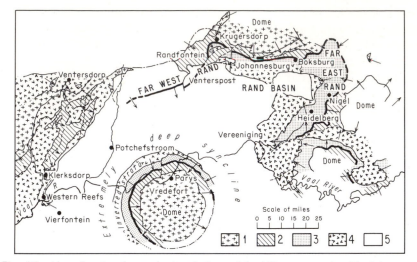

Fig. 47. Map showing the geologic relations of the Witwatersrand gold mining area of South Africa. (1) Old granite, (2) Lower Witwatersrand, (3) Upper Witwatersrand, (4) Ventersdorp, (5) Transvaal and Karroo. Thick solid line denotes Main Reef group, and broken lines are faults. (From DuToit, *Econ. Geol.*, Vol. 35, 1940.)

lower unit. Near the bottom of the Upper Witwatersrand are four gold-bearing conglomerate beds called the Main Reef group. They are named Main Reef, Main Reef Leader, Middle Reef, and South Reef. These beds vary in thickness between 1 and 14 feet, and they are rarely all present in the same area. The Main Reef rocks are the richest gold-bearing units. The Bird Reef group is another sequence of conglomerates that lies about 2000 feet above the Main Reef group. These contain gold but are only locally rich enough to mine. Lying on the Bird Reef group are up to 600 feet of shale above which are the Kimberley reefs, conglomerates that are best developed and worked for their gold content in the West Rand. Thick quartzites including the nonproductive Elsburg reefs make up the rest of the Upper Witwatersrand. About 5000 feet of basic lava flows, called the Ventersdorp System lie directly on the Witwatersrand.

All of these rocks were folded into a synclinal structure and were then eroded to a nearly level surface. Many thousands of feet of sedimentary rocks of the Transvaal System were then deposited, after which a period of igneous activity resulted in the intrusion of many dikes and larger bodies. South of the syncline are three small granite stocks, and the great Bushveld Complex was injected at this time only a few score miles to the north of the area. Granitic dikes that cut the Bushveld also extend into the Rand area. Other periods of erosion dissected these intrusions, and more recent sedimentary rocks now partially cover the older structures and igneous rocks.

The ore bodies are mostly confined to the Main Reef group of conglomerates that can be traced with remarkable continuity around the north and east flanks of the Rand syncline. The richest gold ores occur in what appear to be old channels in the conglomerates where the pebbles are larger and better sorted. These "pay streaks" lie in the reef and commonly pitch diagonally downward to the southeast. The gold-bearing conglomerate is called *banket,* the Dutch word for "almond cake," a pastry that some of the ore is said to resemble. Gold is present as minute flakes and very irregular particles and seems to be mostly confined to the matrix that cements the quartz pebbles. Hydrothermal solutions have penetrated the conglomerates as evidenced by the presence of such secondary minerals as chlorite, sericite, pyrite, tourmaline, chalcopyrite, stibnite, sphalerite, and pyrrhotite. Some true nuggets of osmiridium[1] are present, but rounded nugget forms of gold are almost completely lacking. It has recently been discovered that the banket contains significant and recoverable quantities of uranium present as uraninite that is closely associated with graphite and the gold. Silver occurs as an impurity in the gold in which it averages about 10%.

The origin of the gold in the banket deposits is still contested by geologists, and two theories have evolved, each with a wealth of evidence and a strong following of scientists to support it.

The oldest theory, still strongly advocated by the geologists of the mining companies and government, proposes that the banket ores are ancient placer deposits, the gold having been derived from the erosion of older deposits that still lie to the west and northwest of the Rand. The concentrations of gold in the channels and the widespread and continuous distribution of gold throughout many miles of reefs are facts most convincingly explained by the placer theory of origin. The absence of water-worn nuggets of gold is rather lamely accounted for, however, by the suggestion that the placer gold was dissolved, redeposited, and recrystallized by the hydrothermal solutions that obviously affected these conglomerates.

---

[1] A metallic mineral that is a mixture of osmium and iridium and commonly containing other platinum-group metals.

The alternate theory is one that has been supported mainly by American geologists who have studied the area. They contend that the gold was introduced by hydrothermal solutions. It was deposited in the permeable conglomerates, with extra high concentrations developing in the "pay streak" channels where the solutions moved with greater ease. This theory explains most of the facts but meets with difficulties in accounting for the very widespread and remarkably persistent gold content in the Rand conglomerates. Adequate sources for hydrothermal solutions in sufficient quantities to affect so great an area are not apparent in the region. Of course, this problem affects both theories, each of which calls upon hydrothermal solutions.

The origin of the uranium in the ore has added a complication that must be answered by the above theories. Most proponents of the hydrothermal theory just point to the uraninite as simply another mineral deposited by these hot solutions. Miholić[2] has recently suggested that the uranium was concentrated from sea water by organisms that grew in deep basins in which the gravels were accumulating from a nearby steep coast. This theory accounts for the carbon that is almost always present with the uranium as well as the seepage of methane gas that has polluted the air in some of the gold mines. The gas is thought to be generated by the radioactivity of the uranium acting on the carbon. Carbon is considered the precipitant for the gold which was introduced by later hydrothermal solutions.

SIERRA NEVADA PLACER DEPOSITS

These deposits of California make up one of the greatest placer gold fields ever discovered. They occupy an area about 150 miles long by 50 miles wide on the lower and middle slopes of the western side of the Sierra Nevada Mountains (Fig. 48). The approximate center of this belt lies just west of Sacramento, California. Since their discovery in 1848, these gravels have yielded 1.3 billion dollars worth of gold from ores that ran between $1 and $30 per cubic yard. At the present time, dredges are realizing a profit from gravels which contain as little as $0.10 worth of gold per cubic yard.

---

[2] Miholić, S., Genesis of the Witwatersrand gold-uranium deposits, _Econ. Geol.,_ Vol. 49, pp. 537–540, 1954.

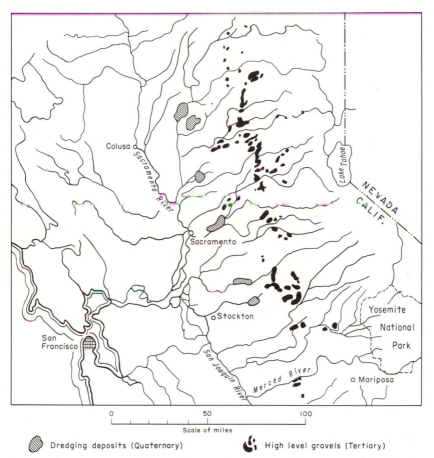

⬡ Dredging deposits (Quaternary)		⬡ High level gravels (Tertiary)	

Fig. 48. Map showing locations of placer gold deposits of California. (After Pardee, *Ore Deposits of the Western States*, A.I.M.E., p. 424, 1933. Compiled from reports of the California State Mining Bureau.)

The source of the gold in these deposits was the Mother Lode vein system that extends for 120 miles in a strip little more than a mile wide. The veins trend northwestward and cut through the slates and other old rocks which lie just to the west of the Sierra Nevada batholith. The Mother Lode has produced over 250 million dollars in gold but is only the remnants of veins which must have penetrated thousands of feet of rock now weathered away. The eroded veins provided the gold, which became concentrated during the Tertiary in numerous placer deposits by streams draining the mountains at that time. At some later time, lava flows covered many of these gravels, and a renewed uplift of the mountains caused the streams to cut deeply into older gravel deposits and form new ones farther down slope. All these Tertiary events culminated in a period of glaciation, but the ice fortunately did not extend far enough down the mountain slopes to erode the gold placers. Although there are many rich deposits in present-day streams, the richest are the Tertiary placers now found high up on the valley walls where they are covered and protected by younger gravels and lava flows.

Silver

Silver, like gold, is one of the precious metals that has been sought avidly through all of historic time. However, most of the production of silver has come as a by-product from the mining of copper, lead, zinc, and gold ores. For this reason the output of silver has not reflected the wide fluctuation in prices between a high of $1.29 and a low of $0.24 per ounce in 1932.

Uses for Silver

About 70% of the world's production of silver is utilized for monetary purposes, either as coinage or as bullion to guarantee the value of paper currency. Much is used for sterling and plated table ware and for jewelry. Great quantities of silver are consumed each year in the manufacture of photographic films and papers. Silver finds minor applications for chemicals, solder, and certain electrical uses.

Production

In 1956 the world production of silver was 230 million fine ounces. Of this, nearly 69% came from mines in the western parts of North and South America. Mexico was the leading producer with 44,000,000 ounces followed by the United States with 38,900,000 ounces. Canada and Peru occupy third and fourth positions with 27,500,000 and 23,000,000 ounces respectively. Australian mines yielded over 14,000,000 ounces. Other countries which had significant silver production were Bolivia, Japan, Germany, Belgium Congo, and Yugoslavia.

Geology of Silver Deposits

Many of the most important deposits of silver have been described as major producers of other metals with silver as a by-product. Over 75% of the United States' silver came from the following mining districts listed in order of their yield: Coeur d'Alene, Idaho; Butte, Montana; Bingham, Utah; Bisbee, Arizona; and Park City, Utah. Of these, the Coeur d'Alene and Park City districts are principally lead and zinc districts, and the other three are noted for their yield of copper. The mining areas described below are not the greatest producers of silver, but they are typical deposits in which the silver in the ores provides the greatest value.

Ore minerals of silver. There are over 50 recognized mineral species that contain silver, but only six of these are important ore minerals. Of course, this list does not include those minerals such as native gold and galena that commonly contain recoverable silver as an impurity. The native silver and cerargyrite are common minerals in the zone of oxidation of a silver deposit, and argentite is usual in the zone of supergene enrichment, although it may be primary. All the minerals containing arsenic and antimony are primary and are deposited by hydrothermal solutions. Native silver may also be primary.

Mineral	% silver	Composition
native silver	100	Ag
argentite	87.1	Ag_2S
cerargyrite	75.3	$AgCl$
polybasite	75.6	$Ag_2S \cdot Sb_2S_3$
proustite	65.4	Ag_3AsS_3
pyrargyrite	59.9	Ag_3SbS_3

PACHUCA DISTRICT, HIDALGO, MEXICO

Mexico has long been the leading silver-producing country in the world, a position gained by the prodigious yields of such rich deposits as occur in the Pachuca district, Hidalgo. This district lies at the southwest base of the Sierra de Pachuca, near the cities of Real del Monte and Pachuca about 60 miles northeast of Mexico City.

The district has had a long and productive history and much valuable ore still remains. The Spaniards were mining in the area as early as 1530, and a British company took over in 1824. Between 1848 and 1906 the mines were operated by a Mexican company that sold out to the American interests who own the deposits at the present time. Water had always been a problem in the mines, and drainage tunnels were driven as early as 1749. Modern equipment and accelerated operations have made the last 50 years the most productive period of the district. The total production of the Pachuca district is estimated at over 1¼ billion ounces of silver and 4½ million ounces of gold.

The rocks of Sierra de Pachuca and the mining area (Fig. 49) are Tertiary lava flows and pyroclastics that have accumulated to a thickness probably in excess of 6000 feet. Andesite is the oldest, thickest, and most prominent rock type of the district, and in it are nearly all the ore bodies. Lying on the andesite are limited flows of rhyolite, and a still younger dacite makes up most of the mountain mass to the north. Basalt is the youngest volcanic rock and is present only in limited patches mostly in the lower southern parts of the district. Dikes and irregular stocklike masses of rhyolite and dacite intrude the older rocks.

There are about 70 veins in the area which have formed in three dominant fault systems: (1) east-west to northwest-southeast, (2)

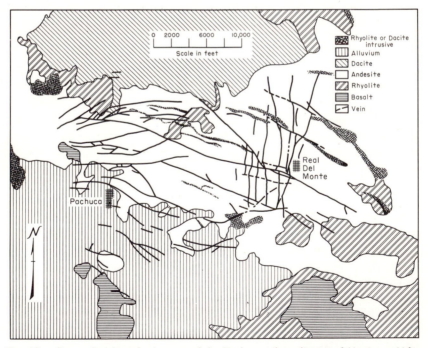

Fig. 49. Generalized geologic map of the Pachuca silver district of Mexico. (After Thornburg, *Min. Eng.*, Vol. 4, p. 596, 1952.)

north-south, and (3) northeast-southwest. The east-west system of fractures was first to form and developed on the north limb of a major northwest anticline. The fracturing was accompanied by formation of dikes that filled many of the fissures. Strong horizontal movements took place along the fractures with the north side moving westward, and about this time rhyolite and dacite flows issued from well-defined vents. The north-south fractures in the Real del Monte area opened up as a result of this horizontal movement, and ore solutions entered at the same time these fissures were forming. The northeasterly trending faults also had horizontal displacements with the southeast side moving westward. All of the fault systems are mineralized.

The veins are mostly 3 to 5 feet wide and rarely extend to depths greater than 1800 feet. Quartz is the common gangue mineral, but calcite and rhodonite ($MnSiO_3$) are locally abundant, the latter in the north-south veins. Argentite is the chief ore mineral, and polybasite and stephanite occur in minor amounts. Galena, sphalerite, and pyrite are locally abundant, enough so in some veins that base metals are produced as by-products. Some gold is recovered from the ore.

The andesite in which all the veins occur has been extensively chloritized throughout the whole district by the hydrothermal solutions. The formation of sericite and kaolinite and the introduction of pyrite are also common but more restricted to the vicinity of productive veins. The introduction of chert and the silicification of the andesite rarely extend more than 10 feet away from the vein contacts. Much of the vein material occurs as cement of the highly altered fragments of an andesite breccia. Wall rock immediately adjacent to the veins is taken during mining, for it contains some precious metal. Alteration of the andesite has proven a guide to the location of new ore bodies. Some ore bodies whose upper limits are more than 200 feet below the surface have been found because of the telltale traces of weathered pyritized zones in the overlying rock. Complete weathering rarely extends more than 100 feet down, but oxidation in some veins has been detected at far greater depths. In the oxidized ore much silver is found in the native state. No supergene enrichment has occurred in the deposits.

COMSTOCK LODE, NEVADA

The Comstock Lode was discovered in the 1850's by prospectors from the California gold fields. It lay only 20 miles from the California border along the eastern side of the Virginia Mountains, an offshoot range of the Sierra Nevadas. The district was named after Henry Comstock who was said to be a lazy claim jumper fortunate enough to gain title to the richest part of the lode. The district reached its peak production in the 1870's at which time as many as 1500 miners were employed. The mines have been closed for many years after having yielded over 400 million dollars worth of gold and silver, of which 60% was silver.

The ore occupied a northeast-trending fault, dipping about 45° to the southeast. There has been about 3000 feet vertical displacement. The downthrown side of the fault consists of a thick sequence of volcanic rocks which now rests against a mass of diorite on the other side of the fault. The ore body had a horizontal extent of 2½ miles and was followed downward over 3000 feet. It was locally several hundred feet thick but branched commonly to several smaller veins. Many small fissures and chambers extending upward from the main fault zone contained rich ores. The hydrothermal ore solutions must have risen into these openings sometime during the latter part of the Tertiary and possibly were active even to the present time. The miners were continually plagued by very hot sulfate waters which frequently burst loose and flooded many levels of the mines. The rock temperatures in the lower levels reached 114°F, and the mine waters were as hot as 170°F. Mining conditions must have been deplorable, for it is reported that standby crews were needed to spell workers after an hour or so work in the deeper levels where the air was like that of a Turkish bath. However, miners received liberal wages of $4.00 per day.

Milky vein quartz and crushed volcanic rock composed the gangue material in the ore. Gold, electrum (gold with up to 40% silver), argentite, polybasite, pyrargyrite, galena, sphalerite, chalcopyrite, and some rarer sulfides made up the ore minerals. Much of the ore down to depths of 2000 feet was amazingly rich, and for this reason the deposit has always been a prime example of a bonanza. A distinct zone of supergene enrichment could be recognized at about 500 feet where a marked increase in the argentite content of the ore occurred. The mining operations stopped at 3000 feet because of the dwindling ore and the difficultly controlled mine waters.

TINTIC STANDARD MINE, UTAH

The first discoveries in the Tintic district were made in 1869, and the district gradually expanded for 30 years with the disclosure and production from many new deposits. The older mines are all located on the western slopes of the East Tintic Mountains about 65 miles due south of Salt Lake City. In 1907 after over 100 million dollars worth of silver, gold, lead, and copper had been produced, the Tintic Standard No. 1 shaft was sunk more than 2 miles east of any of the older mines. The workings were in lean ore and were impeded by heat and gas, but in 1916 a winze[3] from the 1000-foot level opened up one of the largest silver and lead ore bodies in the West. This ore body known as the "Tintic Standard Pothole" (Fig. 50) has already yielded over 85 million dollars worth of silver and lead. The whole Tintic district is probably well over the 400 million dollar mark.

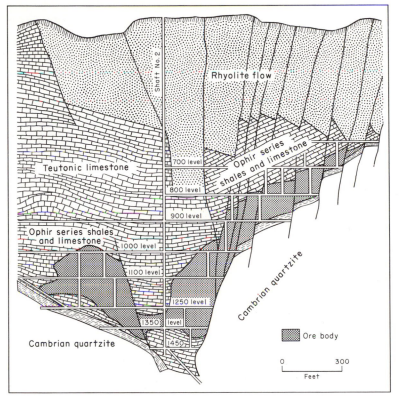

Fig. 50. Cross section trending northeast through the Tintic Standard Mine, Tintic district, Utah. (After Wade, U. S. Bur. Mines Inf. Circ. 6360, 1930.)

---

[3] A winze is a vertical or inclined excavation connecting two levels of a mine. It differs from a *raise* in that it is opened from above.

The geology of the Tintic Standard area is very complex. Thousands of feet of Paleozoic quartzites, shales, and limestones have been strongly folded into an anticline trending northeastward. A northwest-trending fault system and a north-south synclinal crumpling have cut across the crest of the anticline, dropping a portion of the limestones as a funnel-shaped segment into the underlying quartzites. Accompanying the intrusion of several granitic masses, great quantities of hydrothermal solutions were liberated. The liquids utilized this highly disturbed area and rose to the level of the lower beds of limestone where they spread out forming a large replacement ore body. A thick lava flow of rhyolite did not receive any of the metals from the ore solutions but was highly altered by them. It was this zone of alteration at the surface which prompted the exploratory shaft that lead to the discovery of this completely concealed deposit. Post-ore gravity faulting dropped the center of the ore mass even farther, emphasizing the V-shaped character of the body. This settling, thought to be due to shrinkage resulting from replacement of the limestones by the ores and also to oxidation of the ores, resulted in a slight depression in the surface over each of the ore bodies in the East Tintic district. Oxidation of the ore is complete down to the 900 foot level of the mine. There has been little supergene enrichment of this ore body, but a marked concentration of silver occurs in the oxidized parts which are nearly twice as rich in silver as the unaltered material.

The mineralization occurred in two stages. In the lower reaches of the veins and replacement ore bodies the early fluids deposited quartz and barite ($BaSO_4$) with minor amounts of several metallic sulfides. The later solutions extended farther upward replacing the limestone and depositing quartz, galena (silver-bearing), and pyrite. The presence of barite and marcasite (FeS_2) in the ore suggest that the solutions were low temperature.

SELECTED REFERENCES

GOLD

Cooke, H. C., and W. A. Johnson. _Gold Occurrences of Canada,_ Can. Geol. Surv. Econ. Geol. Ser. No. 10, 1932.

Davidson, C. F. The origin of the gold-uranium ores of the Witwatersrand, _Min. Mag.,_ Vol. 88, pp. 73–85, 1953.

DuToit, A. L. Developments on and around the Witwatersrand, _Econ. Geol.,_ Vol. 35, pp. 98–108, 1940.

Emmons, W. H. _Gold Deposits of the World._ McGraw-Hill Book Co., New York, 1937.

Furse, G. D. Geology of the Pearl Lake section of the Porcupine gold area, *Can. Min. and Met. Bull.,* Vol. 47, No. 503, pp. 197–201, 1954.

Graton, L. C. Hydrothermal origin of the Rand gold deposits, *Econ. Geol.,* Vol. 25, Supplement, pp. 1–185, 1930.

Hogg, N. The Porcupine gold area, *Can. Min. Journ.,* Vol. 71, No. 11, pp. 102–106, 1950.

McLaughlin, D. H. Geology of the Homestake Mine, Black Hills, South Dakota, *Eng. and Min. Journ.,* Vol. 132, pp. 324–329, 1931.

McLaughlin, D. H. The Homestake Mine, *Can. Min. Journ.,* Vol. 70, pp. 49–53, 1949.

Moore, E. S. The structural history of the Porcupine gold area, Ontario, *Roy Soc. Can. Tr.,* Vol. 47, Sec. 4, pp. 39–53, 1953.

Noble, J. A. Ore mineralization in the Homestake gold mine, Lead, South Dakota, *Geol. Soc. Am. Bull.,* Vol. 61, pp. 221–252, 1950.

Noble, J. A., J. O. Harder, and A. L. Slaughter. Structure of a part of the northern Black Hills and the Homestake Mine, Lead, South Dakota, *Geol. Soc. Amer. Bull.,* 60, pp. 321–352, 1949.

Sharpe, J. W. N. The economic auriferous bankets of the upper Witwatersrand beds and their relationship to sedimentation features, *Geol. Soc. South Africa Tr.,* Vol. 52, pp. 265–300, 321–330, 1950.

SILVER

Bastin, E. S. Comstock lode, *U. S. Geol. Surv. Bull.* 735, pp. 41–63, 1923.

Billingsly, P. The Tintic District, *XVI Int. Geol. Cong. Guidebook* 17, pp. 101–124, 1933.

Lindgren, W., G. F. Laughlin, and V. C. Heikes. Tintic District, *U. S. Geol. Surv. Prof. Paper* 107, pp. 1–282, 1919.

Thornburg, C. L. The Surface expression of veins in the Pachuca silver district of Mexico, *Min. Eng.,* Vol. 4, No. 6, pp. 594–600, 1952.

Wisser, E. *Pachuca Silver District, Mexico,* A.I.M.E. Tech. Pub. 753, 1937.

Wisser, E. Tectonic analysis of a mining district, Pachuca, Mexico, *Econ. Geol.,* Vol. 46, No. 5, pp. 459–477, 1951.

nickel, chromium, and platinum

Nickel

Uses for Nickel

Very little pure nickel is used by industry except small amounts needed for plating. More than 95% of the nickel is alloyed with other metals, particularly with iron. Steels containing nickel have high tensile strength, great elasticity, superior hardness, and resistance to shock, wear, and corrosion. They are in great demand for all forms of armament, for aircraft and for machinery in industrial plants. There are many other alloys and uses for nickel. The five-cent coin is an alloy of nickel and copper. *Monel metal* (67% nickel, 28% copper, 5% iron) is highly resistant to salt water and is used for pumps, marine propellers, mine screens, and many other such products. *Permalloy* with 80% nickel and 20% iron is widely used for all kinds of electromagnets because it will magnetize and demagnetize so rapidly. *Alnico,* an aluminum-nickel-cobalt steel is used to make the very strongest permanent magnets. There is a great variety of other special nickel alloys.

Production of Nickel

The most important source for nickel at the present time is the great Sudbury district in Canada, which in 1955 produced 175,173 short tons of the metal. This amounted to over 75% of the world's production of 225,271 tons. In this same year the island of New Caledonia produced 27,000 tons, Cuba 15,138 tons, and the Union of South Africa 2,598 tons. The United States' production of 4862 tons was the by-product of copper refining.

Geology of Nickel

Nickel, as well as chromium and platinum, is almost universally associated with the ultrabasic igneous rocks and their alteration products. Most deposits of these metals had long been classified as magmatic, but in recent years good evidence has been discovered suggesting a hydrothermal origin for many of the occurrences of nickel and chromium.

Ore minerals. *Pentlandite* (Fe,Ni)S is the most common ore mineral of nickel. *Millerite* NiS, *niccolite* NiAs, and *nickeliferous pyrrhotite* are other primary nickel minerals. These sulfides and arsenides weather to the two green oxidation minerals of nickel, *annabergite* $Ni_3As_2O_8 \cdot 8H_2O$ and *garnierite* $(Ni,Mg)SiO_3 \cdot nH_2O$.

SUDBURY DISTRICT, ONTARIO

This, the greatest nickel-producing area in the world, is located about 250 miles east of the Sault Sainte Marie in eastern Ontario. The deposits lie around the edge of an elliptical area 36 miles long and 26 miles wide oriented with the long axis trending northeast.

Nickel was first discovered at Sudbury in 1885, and since 1903 the area has been the major source of this important metal. From the start the Sudbury deposits have been intensely studied by geologists because of their economic importance, their unique geologic setting, and their appointed role as the classic example of one kind of magmatic ore deposits. Consequently the literature on the area is voluminous and controversial.

All the rocks of the area are Precambrian, ranging in age between Keewatin and Upper Huronian (Fig. 51). The most striking feature is the great norite[1] intrusion which now occurs as a basinlike structure with its elliptical ring of outcrop. It shows a remarkable gradation from the basic norite at the bottom to a light-colored granite near the top, a feature explained by magmatic differentiation and the alteration effects caused by the streaming upwards of volatile components of the magma. Completely surrounded by the norite outcrop and conforming to the basin structure is a sequence of Upper Huronian(?) volcanic tuffs and agglomerates that have been highly altered by the solutions from the norite magma. Surrounding the intrusion and lying below it are many sedimentary and volcanic series, the oldest of which have been classified as Keewatin. All of these rocks have been greatly metamorphosed. Many smaller intrusions have formed as dikes of granite, basalt, and diabase. Dikes and irregular intrusions of quartz diorite occur sometimes in direct contact with the norite and seem to be later than but related to the norite. Nearly all of the ore bodies are in or close to the quartz diorite intrusions. Extensive faulting, both before and after mineralization, complicates further the geologic picture.

As outlined by Yates, there are three characteristic types of nickel deposits: (1) disseminated ore mostly in quartz diorite, (2) massive sulfide ore in definite fault and breccia zones, and (3) sulfide stringers that form intricate patterns through shattered and brecciated rock. Types 2 and 3 occur in country rock of any kind but never far from intrusions of the quartz diorite. The ore minerals are dominantly pyrrhotite, pentlandite, and chalcopyrite, which occur in amounts to give a copper to nickel ratio of about 1:0.6 in the Falconbridge massive ore and about 1:1 in the disseminated ore of the Frood mine. Valuable by-products are recovered as a result of minor amounts of sperrylite ($PtAs_2$), gold, and other metallic minerals. The gangue is largely quartz and calcite and the variety of host rocks for the ores.

The old theory for the origin of the Sudbury nickel-copper ores was that during the process of differentiation the norite magma had separated from it first a molten sulfide liquid which became insoluble in the cooling silicate melt. This metal-rich fluid settled to the bottom of the norite and was injected into the surrounding

---

[1] Norite is a variety of gabbro which does not contain the usual augite but another pyroxene of the enstatite-hypersthene series.

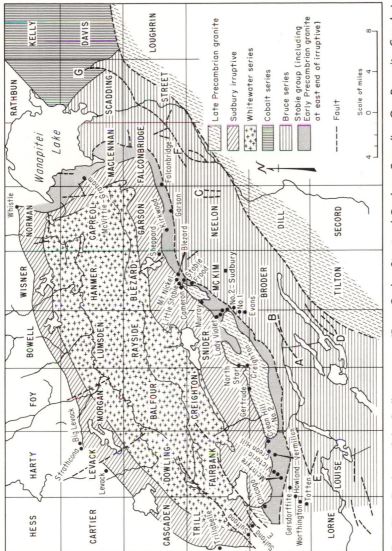

Fig. 51. Geology of the Sudbury area. (From Cooke, *Structural Geology Canadian Ore Deposits*, Can. Inst. Min. and Met., p. 586, 1948.)

rocks to form the ore bodies. Evidence today completely discredits this explanation.

It is now believed that high-temperature hydrothermal solutions deposited the ore, and the movement of these solutions was largely controlled by faults, shear zones, and contacts. It can be shown that the penetration of the ore solutions was the last of a sequence of events including faulting, intrusions of basic dikes, and introduction of the quartz diorite, with which the ore seems genetically related. All these geologic events occurred after the norite had completely solidified. Very little ore of any kind is found in the norite.

NICKEL DEPOSITS OF NEW CALEDONIA

Before the Sudbury district became the major nickel district near the turn of the century, the island of New Caledonia was the world's greatest producer of nickel. It is still the second most important district.

The ores are all secondary, having formed by a lateritization of nickel-bearing peridotite which is exposed over a large part of the island. The peridotite first altered to serpentine on which has formed a weathered mantle of green laterite soil rich in garnierite. There are hundreds of small deposits up to 35 feet thick clustered in six groups, three on each coast. Moving ground water is thought to have played an important role in concentrating nickel, because most of the deposits are on gentle slopes rather than on the level surfaces. The average tenor is 6 to 10% nickel, with 13 to 20% Fe_2O_3, up to 2% cobalt, and 0.1 to 0.8% Cr_2O_3.

Similar though smaller deposits are found in other tropical regions. Cuba, Brazil, Venezuela, and the Celebes have nickel deposits formed by this process of residual concentration.

THE PETSAMO DISTRICT

The Petsamo district lying in that part of Finland annexed by Russia has nickel-copper sulfide ores in an occurrence similar to that of the Sudbury district. The deposit is smaller but contains about 1.6% nickel and 1.3% copper. This area and deposits in the Ural Mountains account for much of the Russian production of nickel.

MYSTERY-MOAK LAKES AREA

This new district in northern Manitoba is being developed by the International Nickel Company and will start shipping nickel in 1960. Nearly 130,000,000 pounds of nickel should be produced each year from the area. An estimated $175,000,000 is being invested to develop the deposits, which will be worked initially from two mines. The ore is closely associated with a peridotite intrusion that was located by an aerial magnetometer survey.

Chromium

Uses for Chromium

Metallurgical requirements take over half of the chromium produced. Chrome steels are exceedingly tough and hard as well as resistant to corrosion, chemical attack, and high-temperature breakdown. The steel alloys containing between 12 and 18% chromium and commonly some nickel are the so-called stainless steels, which have become almost indispensable metals in industries where sanitation is important. Dairies, bakeries, food-processing plants, breweries, distilleries, chemical plants, and many other industries use stainless steels extensively. A special alloy with nickel and iron is called *nichrome* and is used for most of the heating elements in electrical appliances. Chromium with tungsten, cobalt, and molybdenum makes an alloy called *stellite* which is so tough and heat resistant that it is used for high-speed machine tools.

Refractory materials high in chromic oxide and alumina are widely used as basic brick linings for open-hearth and other metallurgical furnaces. Over one-third of the chromite production goes for such purposes. Many brightly colored chromium salts are used by the chemical industry in the manufacture of dyes, pigments, and photographic chemicals. Other chromium chemicals are utilized as bleaches and tanning solutions for hides.

Production

The United States is woefully dependent upon imports to supply our demands for chromite. In 1955 the United States produced

153,253 tons but imported 1,827,960 tons of chromite. Turkey and the Philippines in this year produced 710,000 and 659,310 metric tons respectively. They were followed closely by Russia with 600,000 tons and the Union of South Africa with 597,000 tons. Southern Rhodesia yielded 449,202 tons of chromite, and Yugoslavia, New Caledonia, and Cuba were significant producers. The total world's production of chromite in 1955 was 3,900,000 metric tons.

GEOLOGY OF CHROMIUM

Chromite ($FeCr_2O_4$) is the only ore mineral of chromium. It carries theoretically 68% Cr_2O_3 and 32% FeO, but the oxides of aluminum, iron, magnesium, and calcium are commonly present as impurities. Commercial ores must contain over 45% Cr_2O_3 with a chrome-iron ratio above 2.5:1 to be used for metallurgical purposes. The more impure ores are suitable for refractory chrome.

Nearly all chromite deposits of the world occur in ultrabasic igneous rocks, with which are associated also most of the primary nickel and platinum deposits, and this explains their grouping in this chapter.

SOUTHERN RHODESIA

The important chromite deposits of this country lie in and around the Great Dyke. This striking feature is a dike on truly a grand scale. It extends for 330 miles with an average width of 4 miles, and cuts across all of the broad regional structures along its length. The dike consists of ultrabasic igneous rocks with a layered structure in which the bands dip inward from the edges. The rocks of the Great Dyke are now largely altered to serpentine.

The chromite occurs as persistent bands up to 8 inches thick which are parallel to the banding in the igneous rock. Outside the dike but near its contact with the country rock are chromite lenses up to 450 feet long in talc schist. This region has been very productive of chromite, and much platinum also occurs in certain layers in the dike.

Magmatic segregation is thought by most geologists to be a probable explanation for the banding of the Great Dyke and also the process whereby the chromite was concentrated. It was once believed that chromite was the first mineral to crystallize from the magma, the early crystals settling and accumulating as layers on the bottom of the magma chamber. More recent studies of the Great Dyke have disclosed rather convincing evidence that chromite may also crystallize as one of the last minerals from the residual liquids of the magma and even from the hydrothermal liquids given off from the magma. Studies in other areas confirm this.

RUSSIAN CHROMITE

One of the important chromite deposits of Russia is at Saranovskoye, about 150 miles northwest of Sverdlovsk on the east side of the Ural Mountains. Here schists of Paleozoic age have been intruded by ultrabasic igneous rocks which are now entirely serpentinized. Near the eastern side of the intrusion and enclosed in it are three bands of chromite ore 13 to 30 feet in width. The chromite from this deposit is very low in iron and contains much magnesium. Its formula is best written $(Fe,Mg)Cr_2O_4$. As in the Rhodesian deposits, much of the chromite crystallized after the formation of olivine.

Reserves of ore in this region have been estimated at nearly 7,000,000 tons when the adjacent placer chromite deposits are included. Most of the ore averages only 40 to 46% Cr_2O_3, and much of the lower grade material is used only for refractory and chemical purposes.

Russia has other chromium deposits in South Bashkiriya and in the Kazakh Republic.

UNITED STATES

An interesting but low-grade deposit of chromium lies in the Beartooth Mountains about 40 miles south of Big Timber, Montana. The ores occur in bands averaging about 4 feet thick in a large intrusion of ultrabasic rocks known as the Stillwater Igneous Complex. Although great tonnage is present here, the ore contains only 20 to 27% Cr_2O_3 but can be concentrated to 45%.

Other deposits in the United States are in California, Oregon, Washington, Wyoming, South Carolina, and Texas. All of these are small, and the United States' production of this vital metal falls far short of our industrial needs.

Platinum Metals

The group of six metals, platinum, iridium, rhodium, palladium, osmium, and ruthenium, are similar in properties and usually occur together in nature. Of these, platinum is by far the most common, but it always occurs with such impurities as iron, copper, and one or more of the other metals of the group.

Uses

These metals, although exceedingly rare (platinum is about 100 times as rare as gold), have many important industrial uses which keep them in constant demand. One of their most valuable properties is their ability to act as *catalysts*. This means that the mere presence of one of these metals, usually in a spongy or finely divided state, allows a chemical reaction to take place rapidly. Without the catalyst the reaction would either not go on at all or would proceed at a very slow rate. A catalyst does not take part in a reaction and is not used up. Platinum black (finely powdered platinum) is the catalyst for the important reaction of sulfur dioxide gas with oxygen to form sulfur trioxide, which becomes sulfuric acid when added to water.

The high melting point and chemical inertness make platinum valuable for laboratory crucibles and dishes, for equipment used in the manufacture of explosives, for dental bridges, and as dies for the extrusion of glass fiber and rayon. Its electrical properties and resistance to heat make it useful for heating elements in electric furnaces, for thermocouple wires used in the measurement of high temperatures, for contact points in ignition apparatus such as aircraft magnetos, spark coils, and plugs. Very fine platinum wire is used on tiny overload fuses that blow at a few thousandths of an ampere. The same fine wire may be used for cross hairs in telescopes.

The chemical industries account for over 70% of the domestic consumption of platinum, the electrical industry is second with 13%, and third is jewelry with 9%. Osmium and iridium are used for bearings in fine instruments, as nibs on fountain pens, and even for phonograph needles. The alloys of these metals with each other and also with gold, silver, nickel, copper, and zinc have many important additional uses.

Production

The 1956 records show that the Union of South Africa was the leading producer of platinum-group metals with a yield of 390,000 troy ounces. Canada marketed a total of 385,000 ounces (all metals). Russia's yield came to 100,000 ounces. The United

States and Colombia produced 24,000 and 28,000 ounces respectively. The world total for 1956 was 927,000 ounces. The prices of these metals change constantly, but the following 1956 prices show the comparative values.

1956 AVERAGE COST PER TROY OUNCE

platinum	$107	osmium	$90	rhodium	$122
palladium	24	iridium	105	ruthenium	50

Geology of the Platinum Metals

Ore minerals. Native platinum is the most common ore mineral for all the metals, for it is never found without significant amounts of the other metals as impurities. Natural alloys of the platinum metals include *platiniridium* and *osmiridium,* and others with iron and copper are *fereoplatinum* and *cuproplatinum.* There are several arsenic and antimony compounds containing platinum, but the most common of these minerals is *sperrylite* (PtAs$_2$), a valuable ore mineral at Sudbury.

Examples of ore deposits. The most important producer of the platinum metals at the present time is the great nickel-copper district at Sudbury, Ontario. The rich Frood mine of this district is mining ore which contains an average of about 0.08 ounces of the platinum metals per ton of ore, as well as a significant gold and silver content. The description of the geology of this district is earlier in this chapter.

URAL DEPOSITS

In the north part of the Ural Mountains of Russia there is a sector 375 miles long which is characterized by basic rocks. Within the area are 10 small elliptical masses of ultrabasic rocks which, because of differentiation in the cooling magma, consist of dunite (all olivine) in the center, surrounded by a shell of pyroxenite (rock almost entirely pyroxene), and an outer zone of gabbro. The greatest amounts of platinum are in the dunite, but none of the primary magmatic deposits is rich enough to be mined.

However, erosion of the primary deposits has resulted in the formation of valuable placers in the valleys of streams flowing from these areas. The placers now being exploited are of two ages: (1) the old buried stream valleys, and (2) the more recent streams that cut across the old valleys. Productive sands and gravels are up to 5 feet thick but are often deeply buried by overburden.

Both slopes of the mountains have ore, most of which is mined with great electric dredges.

BUSHVELD, TRANSVAAL

The great Bushveld lopolith north of Pretoria (Fig. 52) contains a layer known as the Merensky Reef which can be traced with interruptions for 300 miles around the outcrop "ring" of the lopolith. The reef is parallel to the layering in the norite, is about 5 feet thick, and dips at an angle of 6 to 25° towards the center of the lopolith. The platinum content is variable, and only in a few places is it high enough to warrant mining. The ore contains platinum, chromite, magnetite, and many sulfides of iron, nickel, and copper. Appreciable palladium occurs with the platinum as well as some rhodium, osmiridium, gold, and silver.

Cutting the layer structure in the Bushveld rocks are many carrot-shaped "pipes" up to 60 feet in diameter and 1000 feet deep. The pipes have concentric zones with a dark dunite core

Fig. 52. (From Wagner, *Econ. Geol.*, Vol. 21, 1926.)

surrounded by a lighter dunite or pyroxenite and on the outside norite. In three of these the central cores are rich in platinum. Assays up to 1200 dwt[2] per ton are reported, but averages of 10 to 20 dwt of platinum are usual. The Mooihoek pipe averages between 6 and 7 dwt. The origin of these pipes is not understood by geologists for the zones seem to indicate a reverse order of mineral crystallization from a magma.

SELECTED REFERENCES

NICKEL

Burrows, A. G., and H. C. Rickaby. *Sudbury nickel field restudied,* Ontario Dept. Mines Rept. 43, Pt. 2, 1934.

Cooke, H. C. Regional structure of the Lake Huron-Sudbury area, *Structural Geology of Canadian Ore Deposits; a symposium,* Can. Inst. Min. and Met., pp. 580–589, 1948.

Davidson, S. Falconbridge mine, *Structural Geology of Canadian Ore Deposits, a symposium,* Can. Inst. Min. and Met., pp. 618–626, 1948.

Lochhead, D. R. A review of the Falconbridge ore deposit, *Econ. Geol.,* Vol. 50, No. 1, pp. 42–50, 1955.

McMillan, W. D., and H. W. Davis. *Nickel-cobalt resources of Cuba,* U. S. Bur. Mines Rept. Inves. 5099, 88 pp., 1955.

Materials survey, Nickel, many unnumbered pp., U. S. Bureau of Mines, Washington, D. C., 1952.

Yates, A. B. Properties of International Nickel Co. of Canada, *Structural Geology of Canadian Ore Deposits, a Symposium,* Can. Inst. Min. and Met., pp. 596–617, 1948.

CHROMIUM

Howland, A. L., R. M. Garrels, and W. R. Jones. Chromite deposits of Boulder River area, Sweetgrass County, Montana, *U. S. Geol. Surv. Bull.* 948C, pp. 63–82, 1949.

Sampson, E. Magmatic chromite deposits of Southern Africa, *Econ. Geol.,* Vol. 27, pp. 113–144, 1932.

Schafer, P. A. Chromite deposits of Montana, *Mont. Bur. Geol. and Mines Mem.* 18, pp. 1–35, 1937.

Zimin, J. A. The Saranovaskoye chromite deposit, *17th Int. Geol. Cong. Uralian Exc. South Part,* pp. 27–33, 1937.

PLATINUM

Buddhue, J. D. Minerals of the Platinum group, *Mineralogist,* Vol. 19, No. 7–8, pp. 350, 352, 1951.

Cochrane, John. The Platinum group metals, *Journ. of Chem. Ed.,* Vol. 31, pp. 407–409, 1954.

Mackay, Robert A. The world's sources of platinum, *Min. Mag.,* Vol. 89, No. 4, pp. 216–218, 1953.

Thomson, A. G. South African Platinum metals, *Min. Journ. London,* Vol. 237, No. 6052, pp. 160–161, 1951.

Wagner, P. A. *Platinum Deposits and Mines of South Africa,* Oliver and Boyd, Edinburgh, 1929.

---

[2] The dwt is the abbreviation of the pennyweight, a troy weight unit equal to ¹⁄₂₀th of an ounce.

uranium and vanadium

chapter 9

Uranium

This metal, long ignored as a rather worthless by-product of vanadium and radium ores, has become one of the most sought-after mineral substances in the world. The great importance of uranium as a source of energy was terribly demonstrated to the world in the awesome blast at Hiroshima on August 6, 1945. In this present time of world-wide tensions, the shadows of the mighty mushroom clouds of the "A-bomb" and the "H-bomb" have acted as dampers on the fires of militarism and agression. For this reason, the thousandfold increase in the production of uranium has gone largely towards the expansion of the nuclear weapons arsenals of the major world powers.

Only a peaceful future can give us largescale industrial harnessing of this vast new source of energy. Already atomic-powered generating plants are operating on a trial basis, nuclear-powered submarines are in operation, and an atomic-powered airplane is said to be not too far in the future. Great advances in medical and other researches have come out of studies made with atomic reactors and materials made radioactive in them. Uranium, the first metal used primarily as a fuel, will certainly be found in more and more places of the world and will be produced in ever greater amounts. Unless an abundant and universal source such as granite can be utilized, however, uranium may never be plentiful enough to supply all the power needs of an industrialized world. The complex equipment needed to utilize this fuel and the hazards entailed in disposing of "ashes" will keep uranium as a specialized source of energy with a favored place (and price) in the economy.

Radioactivity

Uranium is chemically and physically very much like many other metals. It is, however, one of the few elements that constantly emits particles and energy because it is radioactive. To understand this process we must first be familiar with the nature of the atom.

There are three essential particles which make up atoms of all the chemical elements. *Protons* with positive charges and *neutrons* with no charge are of equal weight and comprise the atom's nucleus. The nucleus occupies an exceedingly small part of the volume of the atom yet accounts for nearly all of its weight. Moving at great speed in various spherical levels or shells around the nucleus are *electrons,* particles much smaller than protons, nearly weightless, and negatively charged. Because atoms must be electrically balanced, there are always an equal number of protons and electrons.

The chemical behavior of an element depends upon the number of electrons in its atoms. Formation of compounds with other elements may cause the loss or gain of some electrons, upsetting the electrical balance and changing the atoms into charged *ions.* Chemical changes in no way affect the nucleus, however. The total number of protons, which determines the identity of an element, remains the same.

It is possible for atoms of the same element to have different numbers of neutrons in their nuclei, which causes atoms of different weight. Such atoms that are chemically identical but of different weight are called *isotopes.* Each is distinguished from other isotopes of the same element by its *mass number,* which is the total number of protons and neutrons in the nucleus. Thus, copper with

29 protons may have either 34 or 36 neutrons, making two isotopes with mass numbers 63 and 65. These are written Cu^{63} and Cu^{65}.

Radioactivity is a process whereby the large nuclei of certain heavy elements such as uranium spontaneously break up yielding three kinds of emissions: (1) alpha particles, (2) beta particles, and (3) gamma rays. The alpha particle is positively charged and consists of two protons and two neutrons. Beta particles have negative charges and are electrons which travel at very high velocity. The gamma rays are not particles but are like X-rays with very high penetrating powers. It was discovered that the rate of emission of these particles is unchangeable even by variations of heat, pressure, and chemical combinations of the uranium. Of course, atoms of one element which are losing component parts of their nuclei in this fashion change to atoms of another element. If the new element is also radioactive, it will in turn "decay" to still another element. In this way a natural radioactive series starting with uranium progresses through many different radioactive elements and their isotopes and finally ends as a nonradioactive lead isotope.

Fission occurs when certain heavy nuclei are struck by a rapidly moving neutron and break into two nearly equal halves. A few more nuclear particles are released and an enormous amount of energy. Of the uranium isotopes, only U^{235} undergoes fission. When struck by a neutron its nucleus splits into two lighter atoms such as krypton and barium. At the same time, one, two, or three more neutrons are set loose, and an enormous amount of energy is liberated. If each of the new neutrons encounters a U^{235} nucleus, the collisions are repeated and multiplied. This accelerated process, called a chain reaction, is set off with great rapidity and violence. An unchecked chain reaction is the atomic-bomb explosion. Controlled chain reactions such as in atomic piles may be used to create a steady and usable supply of energy. Unfortunately, only about 0.71% of uranium is U^{235}; most of the rest is U^{238}. The U^{238} is not fissionable and will absorb neutrons without splitting. Therefore, it was once necessary to separate the U^{235} from the U^{238} before a chain reaction could be achieved.[1] It is now a common practice to bombard U^{238} with neutrons and through a series of steps create the new element _plutonium,_ which is fissionable.

---

[1] It has been calculated that the fission of 1 pound of U235 would release energy equivalent to the burning of 1500 tons of coal or 200,000 gallons of gasoline. It equals also the energy liberated in the explosion of 10,000 tons of TNT.

Uses for Uranium

Nearly all of the uranium production today is going into military uses and for scientific investigation (Figs. 53, 54). Some uranium-steel alloys are very hard and tough and may be used for tool steel. Uranium salts are used in the ceramic industry to impart brown, yellow, orange, and even dark-green colors to glass and pottery glaze. Some uranium salts have been used in photography, and many are important chemical reagents in the laboratory.

The unchangeable rate of radioactive decay of uranium has made possible the ingenious use of this metal as a radioactive clock. The physicist has determined that the half life of uranium is 7600 million years. This means that in this period of time 1 pound of uranium would have wasted to 0.5 pound plus many of the daughter products of the decay. The terminal step in this radioactive series is the lead isotope Pb^{206}, which does not disintegrate further. A careful analysis of a fresh unweathered crystal of a uranium-bearing mineral from a granite is made to determine the propor-

Fig. 53. Materials testing reactor at the National Reactor Testing Station in Idaho. (Photo courtesy Idaho Operations Office, Atomic Energy Commission.)

Fig. 54. Technicians loading uranium metal encased in an aluminum tube into the nuclear reactor at Brookhaven National Laboratory, Upton, New York. (Photo courtesy Brookhaven National Laboratory.)

tion of Pb^{206} to U^{238}. By knowing the rate at which the Pb^{206} is produced by the decay of the uranium, one may use this proportion as an indication of the number of years that have elapsed since the crystal formed in the granite magma. The process is difficult and subject to many errors, but several reliable measurements have been made that have given absolute ages to periods of the geologic past. (See Appendix II.)

Production

The production figures and ore reserves of United States' uranium deposits were declassified by the Atomic Energy Commission in December 1956 and released for public use. Data are still very incomplete for world resources and production.

In 1956 the United States processed a total of 6000 tons of U_3O_8 from domestic ores and only 3000 tons in 1957. The following table shows the leading states with their reserves of uranium ores as of 1956.

	Total Reserves	Average % U_3O_8
New Mexico	41,000,000	0.24
Utah	7,500,000	0.34
Colorado	4,100,000	0.33
Arizona	2,600,000	0.30
Wyoming	2,300,000	0.22
Washington	1,500,000	0.18
Others	1,000,000	0.24

Canada announced in 1957 that her ore reserves total 320 million tons averaging 0.12% U_3O_8. France controls ore reserves of 100,000 tons with 0.2% U_3O_8, and South Africa has 1.1 billion tons of 0.03% ore. Australian reserves as of 1957 were 335,000 tons of ore with 0.3% U_3O_8. United States' ore reserves rose to 75,000,000 tons with 0.22% U_3O_8.

Geology of Uranium

The most common primary uranium deposits are hydrothermal veins containing pitchblende associated with silver, copper, cobalt, and nickel minerals. Some pegmatites carry uranium minerals, but only one of these deposits, in Bancroft, Ontario, is rich enough to be mined for uranium alone. Secondary deposits in sandstones are important in the United States, and low-grade occurrences in shale and phosphate rock may soon be exploited. Some placer deposits contain uranium, and commercial placers in Idaho and North and South Carolina are productive now.

Ore minerals. There are over 100 mineral species which contain uranium, but only a few of these have been discovered in concentrations rich enough to be mined. The most important is *pitchblende,* a heavy, noncrystalline, black mineral with a shiny conchoidal fracture. It is essentially uranium oxide, but the composition is variable. Its content of U_3O_8 lies between 50% and 80%. *Uraninite* is a crystalline form of uranium oxide with many of the properties of pitchblende. Pitchblende and uraninite are primary minerals which are found in veins, replacement lodes, and pegmatites deposited from hydrothermal and magmatic solutions.

There are a great number of highly colored secondary uranium minerals which result from weathering or hydrothermal alteration of the primary deposits. The secondary minerals may be deposited far away from their primary source of uranium. *Carnotite,* a potassium uranium vanadate with 50 to 55% U_3O_8, is the most common of these. It is canary yellow and commonly occurs as a cement in sandstones. *Tyuyamunite,* a calcium uranium vanadate, is difficult to distinguish from carnotite, and the two are often associated. *Autunite,* a phosphate of calcium and uranium, occurs as lemon-yellow scales or flakes with a micalike appearance. It fluoresces a brilliant yellow or apple green. Other common secondary minerals are *torbernite, meta-torbernite, uranophane,* and *gummite.*

SHINKOLOBWE MINE, BELGIUM CONGO

This great deposit is located near the southern border of the Congo. The town of Jadotville is nearby, and the city of Elizabethville lies 80 miles to the southeast. The deposits occur in the northern part of the Katanga-Rhodesia copper belt, which extends for about 200 miles in a northwest-southeast direction. The uranium ores were discovered in 1915 during explorations for copper, and production started in 1921. Since that time is has probably produced more uranium and radium than all the rest of the world's deposits combined.

The rocks which contain the uranium ores are units in the ancient "Mines Series." These are mainly dolomites, slates, and quartzites which have been intensely folded and metamorphosed. In many localities the rocks have been thrust faulted in compressive movements that have brought older rocks to lie upon younger ones. The uranium ores occur in one of these overthrusts where a major fault zone intersects it. The small subsidiary fractures have the richest and largest ore bodies, particularly where they cut a thin-bedded quartzite and a dolomitic shale. These beds seemed to have been the most favorable hosts. Also, highly fractured areas have been cemented by ore minerals to produce valuable deposits.

The veins themselves vary in width between a few inches and several feet and seldom extend more than a few feet. However, swarms of such small veins in fracture zones may be mined profitably with all the intervening rock in open-pit type of mining. Many of the veins are pure pitchblende, but as a rule there is a high percentage of cobalt and copper sulfide minerals. The ores yield, besides uranium, important quantities of copper, cobalt, nickel, molybdenum, tungsten, gold, platinum, and palladium. The solutions which deposited the Shinkolobwe ores were probably derived from the same source as those which left the extensive Katanga copper deposits. It is thought that the uranium ores represent a higher temperature deposition, perhaps from the same kind of solutions that formed copper lodes at lower temperatures.

Weathering of the primary uranium ores has been very intense and has extended to great depths. The discovery outcrop was in a small ridge of the more resistant mineralized rock brilliantly colored by the vivid green, yellow, and orange secondary minerals of uranium. Trenching in the area of this ridge revealed some of the rich pitchblende veins just below the surface. In most of the area the depth of weathering is commonly several hundred feet, and little or no pitchblende is to be seen at the surface.

GREAT BEAR LAKE DEPOSITS, CANADA

This important mining district lies just south of the Arctic Circle at Labine Point of Great Bear Lake in the Northwest Territories, Canada. This vast, cold and barren region is dominated by the 200 mile expanse of Great Bear Lake which is surrounded by many rugged granite hills. Steep cliffs mark much of the shore, and the hills rise 1000 feet or more within a few miles of the lake.

In 1900 a Canadian Geological Survey party reported the pink and green staining of copper and cobalt mineralization in cliffs along the east side of McTavish Bay. It was not until 1930, however, that interest was renewed in the area because of a discovery by Gilbert LaBine of rich silver and cobalt veins near the present site of the Eldorado mines. Many more veins were found in the area, including those rich in pitchblende. Now the deposit is of great importance to the atomic industries of both Canada and the United States.

The geology of the area is similar to that of Shinkolobwe. The country rocks are Precambrian metamorphosed tuffs, conglomerates, sandstones, limestones, iron formation, and lava flows. They have been tightly folded, and in the vicinity of the mine their structure is a syncline. These rocks have been intruded by several granite masses around which are zones of intense contact metamorphism. The granites may have been the source of the ore solutions and of the red iron staining which permeates the sediments.

The veins occupy fault zones which now appear in low, shallow valleys carved by glacial ice in the weaker fractured rock. The veins are up to 15 feet in thickness and 700 feet in length. Drilling has shown that they extend to depths over 600 feet. There are four productive veins about 500 feet apart which converge towards the north. The richest masses of ore are found where there had been the most intense fracturing along the faults. Such places occur where faults intersect the more brittle fine-grained metamorphic rocks.

The ores contain more than 40 different metallic minerals. Cobalt-nickel minerals are most prevalent in all the ore bodies, and pitchblende occurs as lenses, large solid masses, or disseminated small particles in the vein fillings. Other valuable ore minerals include native silver, argentite (Ag_2S), chalcopyrite ($FeCuS_2$), galena (PbS), and native bismuth. The gangue in the veins is mostly quartz, siderite, and hematite.

ALGOMA DISTRICT, ONTARIO[2]

The first indication of uranium in this district was in 1847 when J. L. LeConte identified a mineral specimen sent to him from an area north of Sault Sainte Marie. In 1948 a search was started to locate the source of the specimen identified a century before. Many small pitchblende occurrences were found, but none proved to be commercial. However, enough interest was aroused that increased exploration was carried on throughout the area. In 1949 the discovery was made of radioactive conglomerates in the Algoma district only 90 miles east of Sault Sainte Marie. Although large areas gave off a high radioactivity, the assays of surface samples indicated a low uranium content so interest waned. In 1953 the region was restudied, and test drilling revealed ore-grade material at depth. A great rush to stake claims followed, and now there are at least three operating mines in this new uranium district. It promises to be one of the most important in the hemisphere.

The geology of the Algoma region has been studied for over a hundred years, because this is one of the critical areas for the classification of Precambrian rocks. Here and elsewhere along the north shore of Lake Huron those rocks occur that have been named "Huronian," a term also used as a division of the geologic time scale. The Huronian rocks in the Algoma region make a sequence many thousands of feet thick that lies unconformably upon ancient highly metamorphosed granites, gneisses, and other crystalline rocks.

All the Huronian rocks here are sedimentary, and they have been divided into two series. The oldest is the Bruce Series, which has as its lowermost of four formations the Mississagi quartzite, the ore-bearing unit of the area. The Cobalt Series lies uncomformably on the Bruce rocks and is in turn eroded and covered by the basic volcanic rocks of Keweenawan age. All the Huronian and

_____

[2] Originally referred to as the Blind River district.

older rocks have been cut by dikes, sills, and other bodies of Keweenawan basic igneous rocks.

The structure of the region is characterized by two synclines and an intervening anticline, all of which plunge gently towards the west. These folds give a Z-shape configuration to the outcrops of Huronian rocks (Fig. 55). The axis of the south syncline is cut by an east-west major fault along which a mile of horizontal displacement has taken place. Several minor faults trending northwest-southeast cut the north syncline.

The ore bodies lie in the lower part of the Mississagi quartzite in beds and lenses of quartz-pebble conglomerate. Where there are depressions in the pre-Huronian erosion surface, the overlying quartz-pebble conglomerates are thicker and are richer in their uranium content. These depressions occur over the areas of green schist, a rock that was apparently softer and more easily eroded than the surrounding granites. The ore bodies are surprisingly uniform in their uranium content and are very extensive. The lengths of the separate deposits range between 3800 and 7500 feet, and thicknesses vary between 8 and 13 feet. Drilling has proved their

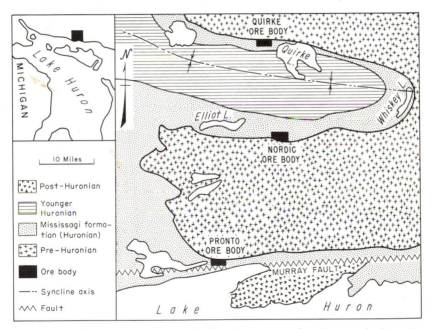

Fig. 55. Sketch map of the Algoma district, Ontario. (After Hart, et al., Can. Min. and Met. Bull. 48, 1955, and Joubin, Can. Min. and Met. Bull. 47, 1954.)

continuity to depths up to 3400 feet. The U_3O_8 content of the ore ranges between 0.1% and 0.125%. Ore proved prior to 1955 had a value of $275,000,000, and it has been estimated that up to 10 times this amount may ultimately come from the district.

Brannerite is the principal ore mineral. It is a complex uranium titanate that contains up to 40% U_3O_8. Most of the brannerite is disseminated through the ore as small rounded or angular grains in the matrix between the quartz pebbles. Some is finely intergrown with the silicate gangue. Grains of pitchblende, uraninite, and monazite are also common. The ore contains abundant pyrite (up to 8%), and the rest of the cementing material of the conglomerate is quartz and sericite (a fine-grained white mica).

The origin of the uranium ores is still debated, and much work is required before any of the theories can be accepted. The great extent of the deposits, their uniform uranium content, the shallow basins on the pre-Huronian erosion surface, and the fragmental nature of the brannerite and other ore minerals all strongly support a sedimentary origin for the deposits. The pyrite and other sulfides suggest a hydrothermal origin for the ore. The Algoma deposits are similar to the gold-uranium deposits of the Witwatersrand, South Africa (see p. 134), and the same theories are proposed for the origin of ores in both districts.

DEPOSITS OF THE COLORADO PLATEAU

This great uranium-producing province is located on the high area of relatively horizontal rocks which is drained by the Colorado River and its tributaries. It is a semiarid region of buttes, canyons, dry drainage areas, and sparse vegetation. The area is often called the "Four-Corners Region" because Utah, Colorado, Arizona, and New Mexico meet at a common point near the middle of the area (Fig. 56).

History. The Indians of the region first made use of the uranium ores several hundred years ago. The bright red and yellow colors made excellent pigment for their war paint. The first discoveries of carnotite were in 1898 in outcrops of the Morrison formation at Rock Creek, Colorado. The ores were first mined in the early twentieth century for their radium content, and for about 10 years during and after World War I they were the world's principal supply. In 1923 the Belgium Congo discoveries resulted in a cessation of all mining here. At this time the vanadium and uranium in the ore were considered of little value. However, the need for vanadium steels created a market which stimulated the mining of

Fig. 56. Index map showing Colorado Plateau. (After Kelley, *U. S. Geol. Surv. Prof. Paper* 300, 1955.)

the uranium-vanadium ores. It wasn't until the utilization of uranium for atomic energy that this metal had more than a by-product importance. Since this time the subsidized market and federal encouragement has drawn thousands of prospectors and mining companies into the area in a mining rush comparable to the California gold rush of 1849. Many great fortunes have been made, and the production of the area has multiplied many times.

Geology of the region. The geologic nature of the region is relatively simple. Permian and Mesozoic sandstones, shales, conglomerates, and limestones which cover the area are essentially flat lying except in many gentle flexures and near a few large faults. The La Sal, Abajo, Henry, and Carrizo mountains are prominent features which stand above the Plateau. They are all igneous intrusive masses, around which the sediments have been folded. Other small igneous plugs, dikes, and lava flows are common, particularly in the southern part of the area.

Carnotite and tyuyamunite are the main ore minerals in many deposits, but uraninite in some areas is of commercial importance. The ore deposits are most common in certain favored sedimentary units: the Morrison and Entrada formations of Jurassic age and the Shinarump and Chinle formations of Triassic age. The ore minerals occur as a cement or coating of the sand grains, as a replacement of carbon, and as staining in fine-grained silts and clays. Ore bodies are commonly flat and conform to the bedding of the enclosing rock. Often a cigar-shaped or wavelike thickening of the ore body is present which has been called a "roll." Some rolls thin out at the edges in a lenslike manner. Few ore bodies are large, but their great number adds up to enormous ore reserves (Fig. 57). The richest uranium ores are a bright yellow, and those with higher vanadium content have a brown color. Many occurrences contain copper and display green and blue colors. In some ores black carbon masks the yellow color.

A few deposits in eastern Utah have little vanadium and no copper. Here the ore mineral is chiefly pitchblende with no colorful secondary minerals, which makes such deposits difficult to distinguish from barren rock.

There are several characteristics of the Colorado Plateau deposits which aid the geologist and prospector in his search for more ore. Following are some of the most useful of these:

Fig. 57. Uranium mine on Colorado Plateau. Miner in foreground is operating a mucking machine; the other miner is working in small stope. (Photo courtesy Union Carbide & Carbon Corp.)

1. Because most of the ores are closely associated with carbon, geologic situations which contain accumulations of fossil wood or asphaltic materials are carefully studied. Some of the earlier discoveries in the area were logs that had been fossilized by uranium minerals and were very valuable. Two petrified logs found in the rocks along the San Miguel River contained $175,000 in radium, $27,300 worth of uranium, and $28,200 in vanadium, all in 105 tons of ore. The largest of these logs was 4 feet in diameter and 100 feet long. In the Calamity Gulch area an ore pocket made up of two logs and the sandstone immediately surrounding them yielded $350,000. These were undoubtedly the most valuable logs ever known.

2. Most deposits are in sandstones and particularly in those beds which are thicker than 20 feet. They are tabular bodies with their long dimensions parallel to the bedding of the sandstone. They almost never extend vertically beyond the limits of the particular host stratum.

3. The silts and shales close to deposits of uranium are usually gray or greenish gray, whereas the same beds elsewhere are red. The sandstones themselves are light brown near the ore bodies and grade to a reddish color in unproductive areas.

Origin. The origin of the uranium deposits of the Colorado Plateau is a problem that has not been satisfactorily solved at the present time. Of the several theories there are two which seem to have the strongest support. One group of geologists believes that the uranium was introduced in hydrothermal solutions derived from the many igneous masses which intrude the area. Instead of making vein deposits, the solutions mixed with ground water, spread over a great area, and deposited their loads where channels of abundant ground-water circulation contained concentrations of carbonaceous material or other conditions that encouraged the precipitation of the dissolved uranium.

The other theory, which seems more plausible to this writer, is that the host rocks and the uranium that was dissolved in the ground water were both derived from the weathering of many hundreds of cubic miles of granitic rocks which took place during Mesozoic times. Such granites normally contain about 4 parts per million[3] of uranium and could easily have supplied enough of this metal to account for all of the deposits on the Colorado Plateau.

---

[3] It is common to report very small percentages in "parts per million." Four parts per million (abbreviated ppm) means in a million grams of rock there are 4 grams of uranium; 1 ppm equals 0.0001%.

The arid climate and internal drainage prevented the loss of the dissolved minerals from the region.

It is notable that whichever of the theories of origin is correct, the factors that control the deposition of these ore deposits are conditions that influence the nature and amount of ground-water circulation. Permeable rocks through which much water has moved can naturally receive more uranium than impermeable rocks. Of course, a "depositing agent" such as carbon is necessary for the formation of valuable concentrations. Excellent examples of the influence of permeability and carbon concentrations are the ancient stream channels at the base of the Shinarump formation (Fig. 58). These old buried lenses of sand, meandering across the eroded surface of the underlying Moenkopi formation, made excellent underground channels for the movement of ground water. As in present-day streams, logs, branches, and other vegetative material accumulated in sharp bends of the Shinarump streams, sometimes as log jams or often merely as logs grounded on sand flats on the inner sides of meander loops. These "trash piles," as the accumulations are called, now lie buried in the sandstones near bends in the old channels. They are the sites of many valuable deposits.

OTHER SOURCES OF URANIUM

There are small percentages of uranium which can be recovered from some phosphate rocks when these materials are treated for

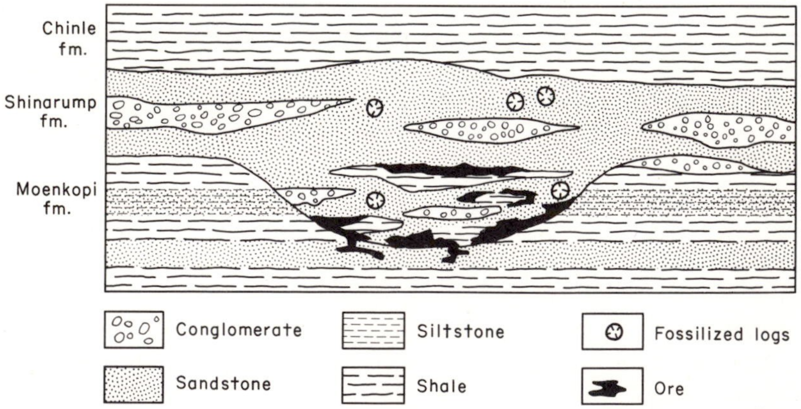

Fig. 58. Schematic cross section of a typical Shinarump channel containing ore. (After Micham and Evensen _Econ. Geol.,_ Vol. 50, p. 172, 1955.)

the manufacture of phosphoric acid and fertilizer. There is a great production potential in the vast reserves of this low-grade material.

A small amount of uranium is present in some petroleum reservoir rocks, and this is often dissolved and brought to the surface in water forced into nearly spent oil pools. One ingenious operator filtered all of the water flushed up from his oil wells through lignite, the brown variety of coal. Just as in the fossil wood in the Morrison rocks of the Colorado Plateau, the uranium deposited under the influence of the carbon. When the lignite was burned the ash was found to contain a recoverable percentage of uranium.

The most promising and abundant possible source of uranium is common granite, which contains about 4 ppm uranium and 12 ppm thorium, which can also be used for nuclear energy. It has been computed[4] that all the energy from these two metals available in 1 ton of granite would be equivalent to that obtained from burning 50 tons of coal. When the energy required to mine and process the granite and extract the uranium and thorium is considered and the proportion of the metals which cannot be recovered is deducted, it was found that 1 ton of granite has available an energy profit equal to that of 10 tons of coal.

Exploration for uranium. Unlike most metals, uranium constantly divulges to us its presence in the rocks and soil if we have the instruments to receive its message. The Geiger-Mueller counter and the highly sensitive scintillation counter react to the gamma radiation of uranium. All prospectors carry such instruments, with which even the most inexperienced can recognize a valuable deposit. The government and larger companies make reconnaissance surveys using counters of very high sensitivity mounted in trucks and airplanes. Maps of radioactivity in a large area made by such methods point to favorable regions for more detailed prospecting. Guided by geologic principles, the prospector thoroughly covers the favored area with his counter, looking for places with exceptional radiation or where the uranium minerals crop out at the surface. After a claim has been staked a drilling program determines how much ore is present, what is its percentage in uranium, and what is the shape and attitude of the ore body in the rocks. However, with only the clicks of their counters to guide them many nonprofessional prospectors have through diligence and luck "struck it rich."

---

[4] Harrison Brown and Leon T. Silver in reference compiled by Page *et al.*

Vanadium

Uses for Vanadium

Pure vanadium has no known uses, but over 90% of the world's production is utilized in the manufacture of vanadium steels. The addition of as little as 0.5% to steel will greatly increase the tensile strength. With also a small amount of chromium in the steel, the properties are just right for springs, tools, forgings, large machine shafts, etc. Some vanadium is alloyed with aluminum and with bronze. The vanadium salts are useful in the processing of leather, in photography, in medicine, and as catalysts.

Production

The United States is the world's foremost producer of vanadium, but, because the production of this metal is closely related to the yield of uranium, a policy of secrecy since 1948 has prevented publication of production data. In 1955 these figures were made available for the first time.

The United States contributed about 89% of the total world output, or 4983 short tons of vanadium. South-West Africa was second with 499. Peru, a leading producer in 1951 and 1952, yielded only 78 tons in 1955.

Geology of Vanadium Ores

Vanadium is actually more abundant in the rocks at the surface of the earth than such metals as copper, zinc, and lead. Despite this fact, it rarely occurs in concentrations rich enough to be considered ore but is almost universally present in minor amounts in basic igneous rocks and many kinds of sedimentary and metamorphic rocks. Although there are over 30 different minerals containing vanadium as a fundamental constituent, most of it occurs as minor impurities in a variety of minerals.

Ore minerals. Of all the minerals of vanadium there are four from which most of the production is achieved. _Carnotite,_ the potassium-uranium vanadate ($K_2O \cdot 2U_2O_3 \cdot V_2O_5 \cdot 3H_2O$), and _roscoelite,_ a vanadium-bearing mica are the principal ore minerals from the Colorado Plateau region. _Vanadinite_ is a beautiful red to yellow mineral with the formula $(PbCl)Pb_4(VO_4)_3$. It is a valuable vanadium-lead ore in Arizona, New Mexico, Rhodesia, and South-West Africa. *Patronite* (VS_4) is the chief ore mineral from the valuable vanadium deposits at Mina Ragra, Peru.

COLORADO PLATEAU

Most of the vanadium produced from this area is recovered as a valuable by-product of the carnotite uranium-vanadium ore. However, for many years when uranium was of little importance the vanadium mica or roscoelite deposits in this area were worked extensively.

The roscoelite ores were discovered near Placerville, Colorado, about 1900 and yielded nearly 4,000,000 pounds of vanadium between 1910 and 1920. After 20 years of inactivity mining was resumed in 1940.

The ore occurs as an impregnation of vanadium minerals in a layer within the upper 25 feet of the flat-lying Entrada sandstone. The ore layer is exposed in two belts, one 10 miles in length and the other 2 miles. The vanadium-bearing bed is continuous in these belts, but only here and there does it become thick enough to mine. These thickenings are in many instances very similar to the so-called rolls which are typical of carnotite deposits. Roscoelite is the main ore mineral and occurs as coatings on the sand grains and partial fillings between them. The color of the ore is greenishgray, ranging to a dark gray in the richer material.

The origin of the ore deposits is not well understood. The area is traversed by numerous gravity faults which displace many of the ore bodies. There are several basic dikes and a stock near the eastern limits. However, neither the igneous activity nor the faulting played a part in the deposition of roscoelite but were in fact younger events. The ore bodies are thought to be deposits located at an ancient ground-water table, but the source of the vanadium and reasons for its deposition are not known.

PERUVIAN DEPOSIT

One of the greatest vanadium deposits in the world occurs at Mina Ragra, near Cerro de Pasco, Peru. This large ore body, 30 feet thick and 350 feet long was formed when a Cretaceous shale, very rich in carbon and asphalt and with many coal seams, was intruded by several dikes. The heat created natural coke, and solutions, apparently from the magma, introduced much sulfur and vanadium. The richest ore contains patronite, pyrite, and red calcium vanadate and occurs most commonly in cracks in the shale, particularly where there is abundant asphalt. A strange mineral substance called *quisqueite* forms the outer zone of the deposit. It is a black lustrous material composed chiefly of carbon and sulfur. The rich ore from the area is sorted and shipped with about 11% V_2O_5. The lower grade ore contains so much carbon, asphalt, and sulfur that it is possible to concentrate a valuable mineral residue by burning it. The vanadium in the ash assays 22% V_2O_5. There are great reserves of low-grade ore in this district.

SELECTED REFERENCES

Bateman, J. D. Recent uranium developments in Ontario, *Econ. Geol.,* Vol. 50, pp. 361–372, 1955.

Bain, G. W. Geology of fissionable materials, *Econ. Geol.,* Vol. 45, pp. 273–323, 1950.

Derriks, J. J., and J. F. Vaes. The Shinkolobwe uranium deposit; current status of our geological and metallogenic knowledge, *International Conference on the Peaceful Use of Atomic Energy, Proceedings,* Vol. 6, pp. 94–128, United Nations Publications A/conf. 8/6, 1956.

Fischer, R. P. Uranium-bearing sandstone deposits of the Colorado Plateau, *Econ. Geol.,* Vol. 45, pp. 1–11, 1950.

Fisher, N. H., and C. J. Sullivan. Uranium exploration by the Bureau of Mineral Resources, Geology and Geophysics, in the Rum Jungle Province, Northern Territory, Australia, *Econ. Geol.,* Vol. 49, pp. 826–836, 1954.

Gruner, J. W. Concentration of uranium in sediments by multiple migration—accretion, *Econ. Geol.,* Vol. 51, No. 6, pp. 495–520, 1956.

Hart, R. C., H. G. Harper, and others. Uranium deposits of the Quirke Lake trough, Algoma District, Ontario, *Can. Min. and Met. Bull.,* Vol. 48, No. 517, pp. 260–265 (*Can. Inst. Min. and Met. Tr.,* Vol. 58, pp. 126–131) 1955.

Joubin, F. R. Uranium deposits of the Algoma District, Ontario, *Can. Min. and Met. Bull.,* Vol. 47, No. 510, pp. 673–679 (*Can. Inst. Min. and Met. Tr.,* Vol. 57, pp. 431–437) 1954.

Kidd, D. F., and M. H. Haycock. Uranium deposits of Great Bear Lake, Canada, *Geol. Soc. Am. Bull.,* Vol. 46, pp. 879–960, 1935.

Mackin, J. H., and D. L. Schmidt. Placer deposits of radioactive minerals in Valley County, Idaho, *Geol. Soc. Am. Bull.,* Vol. 64, p. 1549, 1953.

McLaren, D. C. Vanadium, *Min. Mag.,* Vol. 71, pp. 203–212, 1944.

Nininger, R. D. *Minerals for Atomic Energy,* D. Van Nostrand Co., New York, 1954.

Page, L. R., H. E. Stocking, and H. B. Smith. (compilers) *Contributions to the geology of uranium and thorium by the United States Geological Survey and Atomic Energy Commission for the United Nations International Conference on Peaceful Uses of Atomic Energy,* Geneva, Switzerland, 1955, U. S. Geol. Surv. Prof. Paper 300, 1956.

Proctor, P. D., E. P. Hyatt, and K. C. Bullock. *Uranium, Where It Is and How To Find It,* Eagle Rock Publishers, Salt Lake City, 1954.

Russell, W. L. and S. A. Scherbatskoy. The use of sensitive gamma ray detectors in prospecting, *Econ. Geol.,* Vol. 46, pp. 427–446, 1951.

Thoreau, J. *Concentrations Uraniferes du Katanga* (Shinkolobwe), XVI Int. Geol. Cong., Washington, D. C., 1936.

U. S. Atomic Energy Commission and U. S. Geol. Surv. *Prospecting for Uranium,* U. S. Govt. Printing Office, 1951.

Wright, C. W. Vanadium in Peru, *Foreign Minerals Quart.,* Vol. 3:1 pp. 36–38, 1940.

Wright, R. J. *Prospecting with a Counter,* U. S. Atomic Energy Commission, U. S. Govt. Printing Office, 1954.

miscellaneous metals

chapter 10

Many of the metals discussed in this chapter are fully as important as several of those treated in the preceeding pages, but limitation of space allows only a brief explanation of their nature and geologic occurrences. Some, such as manganese, mercury, and molybdenum, are given favored treatment because of greater utilization, interesting examples of deposits, or both. Others, such as cobalt and cadmium, are described very briefly because these are primarily by-products from ores of other metals which have been discussed more fully.

Manganese

Of all the metals which are used in steel alloys, manganese is the most important, and the one for which there is no substitute. Not only is it used for the high-manganese steels, but it is necessary in common carbon steels where it is added to remove oxygen and sulfur leaving a "clean" metal free from bubble holes. For this purpose each ton of steel requires about 14 pounds of manganese added in the form of *ferromanganese* (80% manganese) or as *spiegeleisen* (20% manganese). The manganese steels, containing over 7% manganese, are very tough, strong, and hard, so they may be used for such purposes as armor plate, projectiles, safes, rock crushers, machine tools, rails, etc. There are important uses for many of the compounds of manganese. The oxide is employed in great amounts for the manufacture of dry-cell batteries and as a decolorizing substance in the glass industry. Other manganese compounds are used in paints, dyes, fertilizers, fluxes, and even medicines.

Russia is the leading producer of manganese, accounting for over one-third of the world's production. The 1956 data show that U.S.S.R. mined 5,235,000 tons of ore, while India produced over 1,824,000 tons of 40% manganese ore. The Union of South Africa and the Gold Coast (Ghana) combined had a yield of 769,000 tons divided nearly equally. The Ghana ore is exceptionally high grade, averaging almost 50% manganese. French Morocco and Cuba have significant yields, but the United States produced only 344,735 tons. She is very dependent upon imports for this strategic metal. The world's production of manganese ore in 1956 was 12,145,000 tons.

Geology of Manganese Deposits

Ore minerals. The most important manganese minerals are the group of oxides: *pyrolusite* (MnO_2), *manganite* ($Mn_2O_3 \cdot H_2O$), *psilomelane* ($MnO \cdot MnO_2 \cdot 2H_2O$), and *hausmannite* ($Mn_3O_4$). They are commonly mixed with one another, and all are black and sooty looking. *Rhodochrosite* ($MnCO_3$) is a rose-colored mineral like calcite and is locally useful as an ore mineral of manganese.

Primary deposits of manganese minerals occur in hydrothermal veins and replacement deposits but rarely in concentrations rich enough to be considered ore. Most of the valuable deposits of the world are classed as sedimentary and residual. Manganese behaves chemically much like iron in the formation of sedimentary rocks, and the two occur together in fresh water lake and bog deposits and in marine sediments. The great Russian ore bodies are in marine sandstones and shales. Just as iron is concentrated by the process of lateritization, manganese also forms residual deposits where manganese-bearing rocks are subjected to tropical weather-ing. Consequently, many of the leading countries in the production of this metal owe their rich deposits to a tropical climate.

RUSSIAN DEPOSITS

Most of the Russian manganese production has come from two great deposits, one in Tchiaturi, Georgia, and the other in Nikopol in the Ukraine. Both are sedimentary deposits which occur in Tertiary rocks.

The Tchiaturi district lies to the south of the Caucasus Mountains about 90 miles to the east of the Black Sea port of Poti. The deposits are in flat-lying Eocene marls, sandstones, and clays which rest upon Cretaceous chalk. The ore horizon is between 6.5 and 10 feet thick and crops out in a region 6 miles wide by 19 miles long. Many canyons cut these strata, and mining is done by means of tunnels through which the ore is removed from under about 300 feet of overburden.

The ore reserves are estimated at over 160 million tons containing about 26% manganese in psilomelane and pyrolusite. One ton of shipping concentrates (50 to 53% manganese) is left from the washing and concentration of up to 3 tons of ore.

The Nikopol district includes two areas totaling about 85 square miles which are underlain by an ore bed averaging about 6.5 feet in thickness and assaying about 30% manganese. The ore consists of lenses and nodules of oölitic pyrolusite, manganite, and iron oxides in Oligocene sandy clay. The deposit is clearly sedimentary in origin, the manganese having precipitated directly from sea water, probably as a result of bacterial and algal activity. The ore reserves are believed to be almost 400 million tons.

INDIA

The Balaghat and other districts of the Central Provinces are the major manganese-producing areas in India. Of slightly lesser importance are the Sandur and Vizagapatnam deposits of Madras. All of these deposits lie in rocks of the Archaean Dharwar System, particularly in the Gondite Series. The Gondite Series contains metamorphosed sedimentary rocks that are believed to have been originally deposited as chemical precipitates rich in manganese oxides. They occur interbedded with metamorphosed shales and sandstones. The weathered and enriched outcrops of these manganiferous metamorphic rocks supply the ore for the Balaghat district and similar deposits in the Central Provinces. In the Sandur area surface alteration of the Dharwar schists has formed concentrations of ore rich in psilomelane and wad (amorphous, earthy manganese oxide). In the Vizagapatnam district is an unusual basic intrusive igneous rock that contains an extraordinary suite of manganese silicate minerals. Intense weathering of outcrops of this rock has caused residual concentrations of manganese oxides to form.

GHANA (GOLD COAST)

In this country there is a band of manganiferous rocks 600 miles long. However, only one deposit, the Nsuta mine, about 34 miles north of the port of Sekondi, is being worked at the present time. Here on a hill extending for about 2.5 miles is a 35 foot thick capping of weathered fragmental ore containing 40% manganese. Beneath this lie the so-called black ores, which are lenticular masses that occur in altered Precambrian rocks. The deposit is worked by power shovels, and the ore is washed until it carries 50 to 53% manganese for shipping. The reserves are probably in excess of 10 milion tons of ore.

There are two schools of thought on the origin of this deposit. One holds that the ores are residual concentrations from the lateritic weathering of a manganiferous bedrock and are, therefore, very shallow. The other view is that the ore was concentrated in Precambrian times and has been only slightly affected by recent lateritization. The second theory implies that there is likely to be a considerable extension of the ore with depth.

Mercury

Mercury, or quicksilver, is the only metal which is liquid at ordinary temperatures.[1] It was named after Mercury, the mythological fleet-footed messenger of the gods, because of its rapid and elusive globules.

Mercury has many important uses as a metal as well as in chemical compounds. Because it is a good conductor of electricity,

---

[1] Gallium is a metal which becomes liquid above 30°C.

mercury is used in switches and other devices. The ancient Greeks discovered that its ability to alloy with gold made it an excellent substance to extract gold from its ores. Much mercury is used for this purpose even today. Mercurial thermometers and dental amalgams are familiar uses. Besides medicines, the chemical compounds find such diverse applications as fungicides, paints, explosives, and dry-cell batteries.

Southwestern Europe regularly furnishes over two-thirds of the world's supply of this liquid metal. Italy and Spain are the leading producers. In 1955 their yields were 53,520 and 45,000 flasks[2] respectively. Mexican mines marketed 29,878 flasks and the Russian production was estimated at 12,000 flasks. Of the 31,500 flasks distilled from United States' ores in 1957, the greatest part came from California, and Nevada, Oregon, Idaho, and Alaska also contributed lesser amounts. Yugoslavia was the only other important producer with 14,591 flasks. The world's total production for 1955 was 196,000 flasks.

Geology of Mercury Deposits

All deposits of mercury seem to have been formed by low-temperature hydrothermal solutions which precipitated ore minerals as cavity fillings in veins, breccias, stockworks, or any other available openings. However, replacement of the country rock by these low-temperature solutions has given rise to extensive ore bodies. Deposits may occur in rocks of any age, but there is a close association with igneous activity of the Late Tertiary. Two mercury deposits are forming today by cinnabar precipitating from thermal waters at Steamboat Springs, Nevada, and Sulphur Bank, California.

The only important ore mineral of mercury is _cinnabar_ (HgS).

ALMADEN, SPAIN

This deposit on the slopes of the Sierra Morena is the greatest and one of the oldest currently operating mines in the world. Although it has been worked almost continuously for over 2000 years, it still contains reserves sufficient to meet the entire world's

---

[2] Mercury is always marketed in flasks, each of which contains 76 pounds and amounts to about two-thirds of a gallon. The price fluctuates greatly but is commonly over $200 per flask.

demands for another 100 years. Pliny, writing in the first century, tells how 10,000 pounds of mercury were brought yearly to Rome from the mines at Almaden, Spain.

The ore occurs in highly folded and faulted rocks of Silurian age. The sequence of slates and quartzites was intruded by quartz porphyry and basic igneous rocks. Three quartzite layers with nearly vertical attitude were partially replaced with cinnabar. The solutions, probably derived from the quartz porphyry intrusion, must have chosen the quartzite layers for the deposition of cinnabar because of higher permeability, favorable chemical nature, or both. The solutions first introduced flaky, fine-grained white mica, called *sericite,* and then the ore minerals replaced the sericite and quartz.

The three lodes are about 1000 feet in length, and the largest is from 26 to 46 feet in thickness. Two of the veins join at a depth of 80 feet. The mine has been worked to depth of 1150 feet where the two remaining lodes are converging. The mercury content of the ore varies greatly between masses and veins of pure cinnabar and low-grade material containing blebs and streaks of ore minerals. The average tenor is about 8% mercury, a very rich ore for such a large deposit.

CALIFORNIA MERCURY DEPOSITS

A section of the Coast Ranges in California extending for over 400 miles contains most of the richest mercury deposits in the United States (Fig. 59). San Francisco, near the center of the belt of deposits, is the principal United States' market for the metal.

The quicksilver produced from this area played an important role in the development of the California gold deposits after the rush of 1849. It was in great demand by the miners for the recovery of the fine gold from the placer gravels.

Perhaps the greatest mercury deposit was the New Almaden mine, discovered and named about 1830 after the famous Spanish lode. Although it did not live up to its namesake, the New Almaden has yielded over 75 million dollars worth of quicksilver. Its production is very minor today.

The ores were deposited under thick clay gouge which marks the faulted contact between rocks of the Franciscan Series and serpentine which is an altered intrusion of peridotite. Rising solutions, impounded beneath the impervious gouge, deposited their mineral loads as irregular fillings and replacements of cinnabar, metacinnabarite, pyrite, quartz, carbonates, and opal. Much

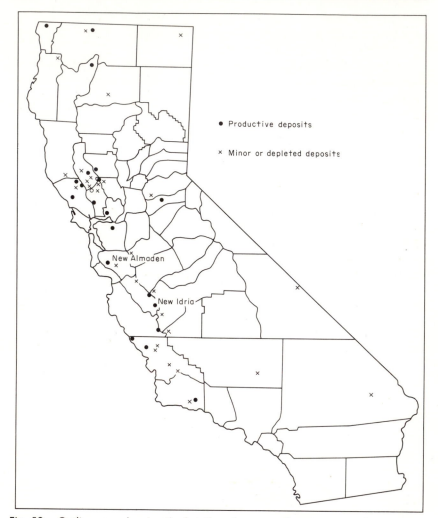

Fig. 59. Outline map showing the location of mercury deposits in California. (After maps in Bulletins 156 (1950) and 78 (1918), Calif. Div. Mines, San Francisco.)

of the ore assayed as high as 12% mercury and was taken from workings as deep as 2100 feet.

The New Idria mine about 70 miles southeast of Hollister, California, is at the present time the leading producer in the United States. The geology here is similar to that at New Almaden with ore occupying two mineralized fault zones between the Franciscan sandstones and a serpentine. This mine, opened in 1850, was named after the ancient and still productive Idria mine in Italy.

Molybdenum

Molybdenum is one of the most important alloy metals in the steel industry. Molybdenum steels are used for ship-propeller shafts, rifle barrels, high-speed tools, and other things where strength and hardness is essential. Because of its high melting point, molybdenum is utilized for electrodes in various kinds of vacuum electronic devices. Some of its compounds are employed in the dyeing industry.

The United States has nearly a monopoly on the world's production of molybdenum. In the first 9 months of 1957 over 44,700,000 pounds of metal were produced, mainly from the mines at Climax, Colorado, and Bingham, Utah. Chile, with 1,300,000 pounds output in the first 6 months of 1957, was the other significant producer in the Free World. It is a remarkable testimony to the increasing demand for this metal that in the period between 1944 and 1948 our average annual production was less than half our present amount.

Geology of Molybdenum Deposits

There is only one important ore mineral of molybdenum, and that is *molybdenite* (MoS_2). The beautiful orange mineral *wulfenite* ($PbMoO_4$) is a minor source.

Molybdenum occurs in deposits of many different geologic types. Pegmatites, contact-metamorphic deposits, hydrothermal replacements, and vein fillings in simple fissures, stockworks, pipes, etc., are all representative. The presence of molybdenite in a deposit is suggestive of high temperatures of formation.

CLIMAX, COLORADO

This deposit is exposed in a glacial cirque on the Continental Divide at an elevation of between 11,000 and 13,600 feet (Fig. 60). It is only 12 miles north of the Leadville district. For many years the large mineralized outcrop attracted attention, but the molybdenite was mistaken for graphite even as late as 1890. Even when it was recognized, many years were required to convince the steel industry of the usefulness of the metal. Also suitable milling procedures had to be developed. The demand for armor-plate steel with the introduction of the tank during World War I was

Fig. 60. Climax, Colorado, site of the nation's largest underground mine which produces most of the free world's molybdenum. The face of the 14,000 foot high mountain at left has gradually receded where molybdenum ore has been removed from beneath. Daily ore production is 35,000 tons; for each ton of ore mined, 6 pounds of molybdenum are extracted. (Photo courtesy Climax Molybdenum Co.)

partially met by the new molybdenum alloys. Mining has been going on at an increasing rate ever since 1917.

The rocks of the district are Precambrian schists and granites which are cut off on the west by the remarkable Mosquito fault. This extensive gravity fault has had a great displacement and now separates Precambrian rocks from the Paleozoic rocks to the west. Tertiary granitic dikes and sills are present on both sides of the fault. In the Precambrian granite quartz veinlets containing molybdenite fill the cracks of a highly fractured stockwork. The ore body has the configuration of a cone nested over a smaller cone. The inner cone is highly altered granite, now almost pure quartz. The ore zones surrounds this as a shell 300 to 500 feet wide and up to 1500 feet in diameter on its inner surface. The ore is surrounded by a zone of slightly altered but intensely fractured granite. At depth the zonation is less distinct, and workable ore bodies exist even in the "barren" quartz core.

The ore is composed mainly of orthoclase, quartz, and sericite in which are disseminated specks of molybdenite and pyrite and small quartz veinlets with sulfides. Minor amounts of topaz, fluorite, sphalerite, and hubnerite ($MnWO_4$) are present but are of no value. The average grade of the ore sent to the mill is 0.83% molybdenite, and enough of this ore is mined to yield about 30 million pounds of molybdenite yearly. The ore reserves have been estimated at over 200 million tons.

BINGHAM DISTRICT, UTAH

This greatest of all copper mines (see page 97) is also the second most important producer of molybdenite. The ore contains only about 0.04% molybdenite which for years was discarded with the mill tailings because it was not considered worth extracting. After it was realized that the mineral could be removed profitably, Bingham quickly jumped into a leading position as a producer of molybdenite. So much ore is processed daily from this mine that the yield of molybdenite nearly equals that from Climax.

Several of the other porphyry-copper deposits of the western states produce molybdenite as a by-product.

Cobalt

Cobalt is another metal used primarily in steel. It causes steels to retain strength and temper even at high temperatures. Such alloys are particularly valuable in high-speed tools and in jet-airplane engines. Over a third of the cobalt is made into magnet steels which are so important in electric motors, speakers, microphones, and other electronic equipment. Chemical compounds of cobalt are used by the ceramic industry as blue coloring agents in glass and enamels. Other applications are as a pigment and a drying agent in paints, as a catalyst, and as part of the nonferrous alloy called "stellite." Cobalt, made radioactive in atomic reactors, is now taking the place of radium for medical purposes. Its intense radioactivity if used in nuclear explosions threatens lasting total destruction for a target area.

In 1957 the total Free World's production of cobalt was estimated at 15,500 short tons. United States contributed only 1848 tons as the second largest producer. The Belgium Congo has the leading mining areas which normally market about 9000 tons yearly. Canada, Northern Rhodesia, and French Morocco were minor producers.

Geologic Occurrence of Cobalt

Cobaltite (CoAsS), *smaltite* (CoAs$_3$), and *linnaeite* (Co$_3$S$_4$) are primary ore minerals of cobalt. The secondary minerals *erythrite* (Co$_3$As$_2$O$_8 \cdot$ 8H$_2$O), with its distinctive pink color, and the black oxide are common minerals of the weathered zone. There is a very close association of cobalt with arsenic in nature.

The Belgium Congo production comes as a by-product from the smelting of the oxidized ores from the great Katanga copper belt. Most of the cobalt is present as the black oxide, which forms as a residuum from the weathering of the same dolomites that contain the oxidized copper deposits. The famous silver mines at Cobalt, Ontario, now mined out, were once major producers of cobalt as a by-product. In the new Blackbird district, Idaho, are high-temperature copper-cobalt veins and replacements in shear zones. Cobaltite and chalcopyrite are the main ore minerals. There is large tonnage indicated.

Titanium

Titanium is the seventh most abundant metal, more plentiful than nearly all of the most common industrial metals except aluminum and iron. However, it rarely occurs in concentrations rich enough to be suitable for ore. It is present in minor amounts in most rocks, and deposits of titaniferous magnetite are common but are seldom used because of the metallurgical difficulties in separating the iron and the titanium.

Metallic titanium is being produced in ever greater amounts during recent years. It is needed for refractory alloys that can withstand the intense heat generated by jet-aircraft engines and rockets. The titanium oxide is still in great demand as the whitest of all white pigments. Its main use is in paint, but it is also employed in the coloring of linoleum, the dyeing of leather and cloth, and in tinting glass, pottery glazes, and even artificial teeth.

Geology of Titanium Ores

Ilmenite (FeTiO$_3$) and *rutile* (TiO$_2$) are the only important ore minerals of titanium. Both minerals are common in igneous and metamorphic rocks, and some magmatic concentrations from basic magmas have created large primary deposits. One of the largest of these is the Lake Allard deposit, Quebec. Here an intrusive body of anorthosite has itself been invaded by what must have been late-magmatic liquids which left nearly pure masses of hematite and ilmenite. The largest ore body has an estimated reserve of over 200 million tons of this rich ore, which contains an average of about 37% titania (titanium oxide).

The Adirondack deposits at Lake Sanford, New York, are similar in occurrence to the Lake Allard ores, late-magmatic concentrations of magnetite and ilmenite in gabbro and anorthosite. Much ilmenite has been produced from the area, and the ores also contain recoverable amounts of vanadium.

Both ilmenite and rutile are moderately heavy minerals which are hard and resistant to physical wear and chemical decay. Consequently, both are concentrated in placer deposits. The beach placers at Trail Ridge, Florida, those south of Jacksonville, Florida, and the extensive black sands at Travancore, India, are such secondary deposits. Ilmenite is easily separated from placer sands by means of electromagnets which attract the slightly magnetic mineral. Nonmagnetic rutile must be removed by other techniques.

United States, India, Norway, and Canada are now the leading producers, but Malaya and Australia have also contributed heavily to the world's supply.

Antimony

Antimony is one of the few metals that expand rather than contract when hardening from the melt. This property and its low melting point make it exceptionally useful in type metal where up to 30% antimony may be alloyed with lead. It also imparts hardness and stiffness to lead, and such alloys are fabricated into pipe, electric-cable sheathing, bullets, solder, foil, and storage-battery plates. A group of important alloys containing antimony are the so-called white metals which are used for antifriction bearings and other purposes. One of these is the familiar *pewter*. Chemical compounds of antimony find application in pigments, fire-proofing compounds, rubber, glass, shrapnel balls, detonating caps, and smoke screens.

Geology of Antimony

Most ore deposits containing this metal were formed by low-temperature hydrothermal solutions at shallow depth. The ore occurs most commonly as fillings in fissures and other rock openings. There seems to be a close natural affinity of antimony for lead, and much of the production is as a by-product from lead ores. The mineral *stibnite* (Sb_2S_3) is the only important ore mineral (Fig. 61) although a great many copper, silver, and lead minerals contain antimony.

Fig. 61. Radiating crystals of stibnite in ore from Manhattan, Nevada.

In China the most productive deposits are in Hunan province near Changsha. Here, in an area 2.5 by 3 miles, deformed and metamorphosed Silurian rocks contain veins, pockets, lenses, and irregular masses of stibnite ore. The richest bodies occur in anticlinal structures. The mine-run ore carries about 6% antimony, but hand sorting brings the shipping ore to as much as 60%.

Bolivian ores come from the 150 mile belt of mineral deposits which are richest near Uncia and Porco. Shallow quartz veins in shale carry stibnite along with the sulfides of other metals. In the United States antimony is mainly a by-product metal from lead ores such as those from the Coeur d'Alene district, Idaho.

Magnesium

The need for lightweight metals for the construction of aircraft and for other purposes has greatly increased the demand for magnesium and its alloys with aluminum. Light from burning magnesium is rich in ultraviolet rays, so that the metal is a main ingredient of flash bulbs. Night flares, sky-rockets, "star shell," "flash bombs," and incendiary bombs also contain magnesium. It is used in metallurgical processes to remove oxygen and sulfur from metal melts. The oxide (magnesia) and other compounds find various commercial applications from insulation material to medical uses. Much magnesia is used in the manufacture of basic refractory bricks for the lining of metallurgical furnaces.

Geology of Magnesium

There are literally hundreds of minerals which contain this common metal, yet most of the production of magnesium is from natural brines and sea water. Magnesium chloride is first extracted from these solutions. This is melted, and the magnesium is separated electrolytically. *Magnesite* ($MgCO_3$), *dolomite* ($CaMg(CO_3)_2$), and *brucite* ($Mg(OH)_2$) are common minerals from which magnesium and, particularly, its oxide are recovered. Sources of magnesium and its compounds are available in unlimited amounts to nearly every country in the world. United States, Germany, Japan, and United Kingdom have been leading producers.

Beryllium

Beryllium, one of the lightest metals known, greatly resembles magnesium and might be called the "new wonder metal" for recently science and industry have found many exciting applications for beryllium.

Most of the beryllium is alloyed with other metals, where it imparts great strength even when added in small amounts. One alloy of copper with about 2% beryllium and about 0.5% cobalt or nickel is now widely used in all kinds of springs because of its exceptional fatigue endurance. The same alloy may be cast into nonsparking, nonmagnetic safety tools. It has been found recently that beryllium adds strength to aluminum-magnesium alloys as well as making them corrosion resistant even at high temperatures. In 1932 physicists bombarded beryllium with alpha particles and found that it gave off a new particle which they named "neutron." In atomic research this metal has long been used as a source of neutrons, and it was also tried as a moderator in early atomic piles. However, for the latter use graphite was found to be just as good and far more abundant. Because beryllium is transparent to the longer wave-length X rays, it has been employed as a "window" in X-ray tubes. Some of its compounds are now used as yellow-white phosphors in fluorescent lamps, and beryllium oxide is considered a super refractory, able to withstand temperatures up to 2570°C.

Unfortunately, the supply of this exciting "new" metal is limited by its restricted geologic occurrence, for in great quantities it could become a major structural material. The world's production of beryl[3] in 1957 was estimated at 7500 tons, mostly from Brazil, Argentina, Southern Rhodesia, Mozambique, and India. United States' mine production was only 460 tons in this year, which is still a marked increase over previous years.

Geologic Occurrence of Beryllium

Beryl ($Be_3Al_2Si_6O_{18}$) (Fig. 62) is the only important source of beryllium, although there are several other minerals which contain this metal. Beryl is sometimes present as a gangue mineral in ores of tin and tungsten. Its principal and only important occurrence, however, is in pegmatites. A good example of pegmatite beryl deposits are those in northeastern Brazil from which over one-third of the world's supply is mined. It is an area of Precambrian schists and quartzites which have been intruded by granite. Hundreds of

---

[3] The mineral beryl contains only 5.04% of beryllium.

Fig. 62. Fragment of beryl crystal showing hexagonal nature.

dikes, lenses, and irregular masses of pegmatitic rock occur in the schists. The largest is 500 by 100 meters. Most of the pegmatite bodies display distinct zones which may be numbered from the outside in as: (1) an outer zone in contact with the schist and rich in coarse white mica; (2) a thick zone of ordinary quartz-feldspar-muscovite pegmatite; (3) a zone of extremely coarse-grained feldspar crystals in which are large beryl, tantalite, and spodumene crystals (this is the ore zone); (4) a central core of rose or milky quartz. The ore is mined from open pits, and the valuable minerals are sorted out by hand. Besides beryl, the tantalum and lithium minerals are marketed.

Bismuth

Bismuth is a shiny white metal which is alloyed with other metals in order to lower their melting points. Wood's metal, for example, is an alloy of bismuth, tin, and cadmium which melts at 60°C. This alloy is widely used in automatic sprinkler systems for fire protection. When the temperature of a room exceeds 60°C, plugs of Wood's metal in the sprinkler heads melt out, starting the water spray. Other alloys find application in molds (like antimony, bismuth expands when crystallizing) electrotypes, antifriction metals, electrical apparatus, and a host of minor uses. Its salts have beneficial medical properties, and because of their soft unctuous feel they are desired for cosmetics. A draught of bismuth salts is taken before X-ray examination of the digestive tract because its opacity to the radiation makes the organs more visible.

Geologic Occurrence of Bismuth

Deposits containing bismuth are mostly those formed by medium- and high-temperature hydrothermal solutions. No deposits are ever mined exclusively for bismuth, but it appears as a valuable by-product in ores of tin, lead, copper, and silver. *Native bismuth, bismuthinite* (Bi_2S_3) and *bismite* (Bi_2O_3) are the main minerals, but most of the United States' production is from lead and copper ores where the bismuth is an impurity in the main ore minerals. It is recovered from the slime left after the electrolytic-refining process.

The United States is the leading producer followed by Mexico, Peru, Korea, and Yugoslavia. The annual world's production seldom exceeds 2000 tons of metal.

Cadmium

Most cadmium is used in alloys with other metals. Wood's metal (see under bismuth) contains 12.5% cadmium. With nickel and copper it makes an excellent antifriction bearing metal for automobiles. Cadmium amalgam is a common dental filling. It is used to harden copper and to make silver resistant to tarnish. "Green gold" owes its color to alloyed cadmium. Steel is sometimes plated with cadmium to prevent it from rusting. Great quantities of cadmium sulfide and cadmium oxide are used as yellow and orange pigments in paint. Other compounds are employed in photography, rubber, soaps, dyes, and ceramic pigment.

Geologic Occurrence of Cadmium

All cadmium is recovered as a by-product from the smelting of zinc ores. Because it is universally associated with this metal and with no others, the countries which lead in the production of zinc are also major producers of cadmium. The only important ore mineral is _greenockite_ (CdS), a yellow substance commonly occurring with sphalerite. When the ore is smelted cadmium is volatilized and is collected with the flue dust. Over 17 million pounds of the metal are produced each year. United States is by far the leading producer, and Mexico, Canada, and Australia contribute important amounts.

SELECTED REFERENCES

MANGANESE

Fox, Cyril S. The mineral resources of Soviet Russia, _Min. Mag.,_ Vol. 83, No. 4, pp. 201–211, 1950.

Kostov, I. The world's manganese ore, _Min. Mag.,_ Vol. 72, pp. 265–270, 1945.

Materials survey, manganese, U. S. Bur. Mines, Washington, D. C., 539 pp., 1952.

MERCURY

Ebner, M. J. A selected bibliography on quicksilver, 1811–1953, _U. S. Geol. Surv. Bull._ 1019–A, 62 pp., 1954.

Schuette, C. N. _Occurrence of quicksilver ore bodies,_ A.I.M.E. Tech. Pub. 335, 1930.

Schuette, C. N. Quicksilver, U. S. Bur. Mines Bull. 335, 1931.

Trengove, R. R. _Investigation of New Idria mercury deposit, San Benito County, California,_ U. S. Bur. Mines, Rept. Inves. 4525, 24 pp., 1949.

MOLYBDENUM

Butler, B. S., and J. W. Vanderwilt. Climax molybdenum property of Colorado, _U. S. Geol. Surv. Bull._ 846, pp. 191–237, 1933.

Hess, F. L. Molybdenum deposits, _U. S. Geol. Surv. Bull._ 761, 1924.

COBALT

Materials survey, cobalt, U. S. Bur. Mines, Washington, D.C., 207 pp., 1952.

Young, R. S.   *Cobalt,* Am. Chem. Soc. Mon. 108, Reinhold Publishing Corp., New York, 181 pp., 1948.

TITANIUM

Barksdale, J.   *Titanium, Its Occurrence, Chemistry, and Technology,* Ronald Press, New York, 1949.

Carpenter, J. R., and G. W. Luttrell.   *Bibliography on titanium (to January, 1950),* U. S. Geol. Surv. Circ. 87, 19 pp., 1951.

Hammond, Paul.   Allard Lake ilmenite deposits, *Econ. Geol.,* Vol. 47, No. 6, pp. 634–649, 1952.

BERYLLIUM

Materials survey, beryllium, U. S. Bureau of Mines, Washington, D. C., 175 pp., 1953.

Antimony, Magnesium, Bismuth, and *Cadmium.*

The General References will add to the brief discussions of these metals given here and will provide additional specific references.

GYPSUM

section two nonmetallic

GYPSUM

ROCK SALT

QUARTZ
CRYSTAL

mineral resources

MICA-TOURMALINE

chapter 11

ground
water

If a dollars-and-cents value could be put upon all of the mineral resources used by man, ground water would probably head the list. But how can we place a specific value on the well water consumed by stock on a western cattle range, the irrigation water spread on a productive sugar beet field in western Nebraska, the water used to air condition a large department store in Minneapolis, or even our individual domestic supply? Water is not marketed like other commodities except at a small cost when supplied by municipal or private distributers. Most people consider it a gift of God, free as the air and always available for those who would drill a hole in the ground and pump it out. For the most part this is true, but failing ground-water supplies in some areas of our country have put a limit to agricultural and industrial growth; legal controversies have been numerous over the exploitation of ground-water reserves; new laws have been passed governing its use; and conservation measures have been recommended for many regions. In short, people are no longer taking for granted this natural endowment. They are learning what a valuable resource is ground water and that it is not inexhaustible.

Uses for Ground Water

More ground water today is used for irrigation than for any other purpose. Countless acres of highly productive farmland in the semiarid regions of America, India, Africa, the Middle East, Southern Europe, and elsewhere would be worthless were it not for the millions of acre-feet[1] of life-giving ground water pumped onto the fields each year. Nearly all kinds of crops may be irrigated, and the combined effect of abundant sunshine and scientific watering result in consistently high yields.

Cities require enormous quantities of water which in many parts of the world can only be supplied by ground-water reserves. Most of the demand is by industrial plants which locate normally in urban areas. Oil refineries, paper mills, chemical plants, breweries, food-processing plants, and many other kinds of industries consume great quantities of water. City sewerage disposal in most areas requires dilution by much water.

Another growing demand for ground water is for air conditioning. It is commonly used even where surface water supplies all other needs, because the ground water is usually colder. Millions of gallons are pumped from wells in the hearts of cities for the purpose of cooling large office buildings, stores, and theaters. The modern-day trend towards home air conditioning is putting an unheard of strain on the capacities of most municipal water systems during the hotter months. According to estimates made by the U. S. Public Health Service, the modern large city of more than 100,000 population has a per capita consumption of water up to 140 gallons per day.

Occurrence of Ground Water

Water in any form that occurs in the ground is called _subsurface water_ (Fig. 63). _Ground water_ may be defined as all uncombined subsurface water in the pores of rocks and soil below the water table. It occupies bedding planes, foliation planes, joints, cavities, and all other openings large and small which are universally present in rocks. The quantity of such water in a given saturated rock is determined by the rock's _porosity_. Ground water may come

---

[1] This is a unit of volume used to measure irrigation water. It is the amount required to cover an acre area 1 foot deep.

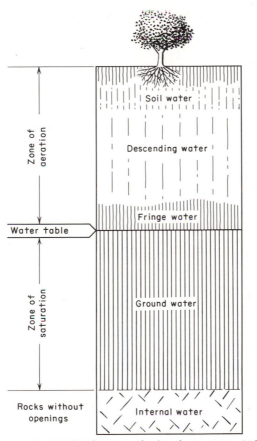

Fig. 63. Diagram showing the divisions of subsurface water. (After Meinzer.)

from three different sources: (1) Magma, crystallizing beneath the earth, liberates great quantities of water vapor which condense upon cooling. This brand-new water is called *juvenile* water. (2) Most sedimentary rocks result from the accumulation of mineral material beneath some body of water. Much water is entrapped in the pores of the newly formed sedimentary rock where it remains as *connate* water. Most connate water is salt water from some ancient ocean. (3) Rain, snow, hail, frost, dew, mist, etc., are all forms of *meteoric* water which have as a source the atmosphere. Nearly all ground water is meteoric water which has soaked into the earth from the surface. Connate water is encountered in deep wells in sedimentary rocks protected from the flushing action of percolating meteoric water. Hot mineral springs in regions of recent volcanic activity are considered to be partially juvenile water, although heated resurgent meteoric water is commonly mixed with it in great amounts. Unless otherwise noted all further references to ground water will imply meteoric water, which is the most important.

Ground-water Table

Nearly everywhere in the world one can dig or drill a hole in the ground and at some depth encounter rock or soil that is saturated with water. Below this depth the openings in the rock are filled with water down to a point where pressures are too high for openings to exist. The upper surface of this zone of saturation is called the *ground-water table*. Above this surface and extending to the surface of the ground the rock and soil openings are filled with air but also contain some moisture that is either moving downward or clinging as film on the mineral grains. Immediately above the ground-water table the extremely fine capillary[2] openings in the rock may be filled upward as much as 10 feet with water that is drawn from the ground-water table by surface-tension forces. This zone is called the *capillary fringe*. It is widest in materials like clays in which the openings are very small and very narrow or nonexistent in materials such as gravel where openings are nearly all above capillary size. The capillary fringe may extend to the surface, and in such circumstances water is constantly released for evaporation. It is important that many crops develop root systems down to this fringe water.

The ground-water table is not a flat surface as its name implies but undulates in such a way that it roughly parallels the surface of the earth. Its hills and valleys lie under surface hills and valleys. However, the ground-water table may reach the surface in deep valleys and lie far below the surface under hills. Thus, the surface topography is reflected by the ground-water table, but with more gentle slopes and less relief (Fig. 64). Wherever there are ponds, lakes, swamps, springs, and ever-flowing streams, the ground-water table has intersected the surface. Many a stream flows above the ground-water table, but only where there is more water entering its channel than is soaking downward into its bed. Bodies of stagnant water quickly dry up if the ground-water table drops below their lowest points.

---

[2] A capillary opening is one so small that the surface-tension attraction of water to its walls is strong enough to overcome gravity, and water is drawn upward above its static level.

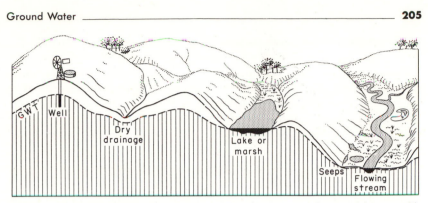

Fig. 64. Schematic diagram showing the relationship between the ground-water table (GWT) and various features of the surface of the earth. A uniform rock or soil material is assumed for the sake of simplicity.

Occasionally in the unsaturated zone above the ground-water table there are lenses of clay or other impervious material which may become a "floor" for a zone of saturation well above and separated from the regional ground-water table. The top surface of this zone is called a *perched water table* (Fig. 65). Ponds and swamps are often caused by perched water near the surface. These bodies may be easily drained by drilling holes in the impervious material, allowing the water to drain down to the normal water table.

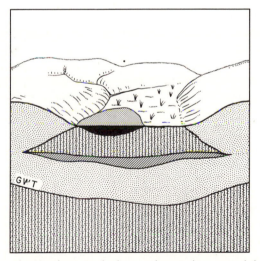

Fig. 65. Diagram showing how perched ground water has caused the formation of a pond and swamp. A lens of impervious clay in the sand acts as the "floor" for the perched water.

Movement of Ground Water

Like water at the surface of the earth, ground water moves in response to the force of gravity, that is, it flows down hill. It is constantly moving downward from high places on the ground-water table to low spots where it may reach the surface in springs and flow off into the surface drainage. The velocity of this subsurface flow depends upon two important factors: (1) the permeability of the rock or soil, and (2) the slope of the ground-water table. Water flows through cavernous limestone with nearly the same ease with which it moves on the surface. However, in most rocks the openings are so small that water seeps through them very slowly. Some rocks, such as dense clays, are nearly impervious to ground water. Those sedimentary rocks which are most permeable have coarse grain size, good sorting, and poor cementation. Open joints, fractures, bedding planes, solution cavities, and other large rock openings often cause high permeability.

The velocity of ground-water movement where the water table is steep will always be greater than where its slope is less in materials of the same permeability. The lower the permeability the slower is this outflow of water from the high points to the low ones, and in more impervious materials a steep slope in the water table can be maintained longer. Of course, were it not for periodic replenishment to the subsurface water by rains and melting snow, the topography of the ground-water table would ultimately be reduced to a featureless plane.

It is difficult to actually measure the rate of ground-water percolation, but computations of velocities based upon permeability and porosity yield values that vary between a low of 1 foot in 10 years and as much 420 feet or more in a day. In most strata and soils from which water is extracted[3] the rate lies between 5 feet per day and 5 feet per year.

Climatic Effect on the Ground-Water Table

After a rain or the melting of snow the level of the water table rises closer to the surface. This is because water moves more rapidly downward through the unsaturated pores above the water table than it is carried off by the slow percolation below the zone of saturation. On the other hand, the surface of the water table

---

[3] Such water-producing strata are called _aquifers_.

gradually lowers at times when there is no water infiltrating from above. Consequently, periods of rainfall or melting snow as well as intervals of drought have their effects on the level of the ground-water table. In regions with definite rainy and dry seasons the water table may fluctuate many tens of feet from its high elevation during the wet time of year to its low elevation during the arid months. The seasonal change in level depends both upon the amount of precipitation which enters the surface and the permeability of the soil.

In desert regions the zone of saturation is commonly several hundred feet below the surface except in deep valleys. There may be little relationship between it and the surface topography. In fact, the highest points on the water table may be under dry stream valleys where downward seepage from occasional flood waters makes a ridge or hump. Such features are only temporary, particularly where the surface drainage channels may receive rain water perhaps only once in a few years.

Confined Ground Water

In the preceding pages we have been concerned with ground water that is free to move in any direction through the pores in the rocks and soil. Near the surface of the earth this is nearly always the case, and a ground-water table results from the balance between subsurface percolation and infiltration from above. Well below the water table may lie a stratum so impervious that water cannot penetrate it. Below this bed may be one with a high permeability which contains water that did not come from above but entered the permeable unit several hundred miles away where it intersects the surface or is not covered by the impervious bed. This water is not free to move in any direction but is confined to the particular bed beneath the impervious layer. If the water-bearing stratum slopes downward from the area where the water enters to where it is removed, wells may gush water under pressure. This pressure or *head* increases the greater the difference in elevation between the intake area and the point of removal.

In Fig. 66 water from the well drilled at point B would spurt as high as plane A–D. In the well at point C the water would also rise to plane A–D. At point B is a flowing artesian[4] well which

---

[4] Artesian wells were named from the flowing wells near Artois, France, and the term now applies to all wells producing from confined aquifers.

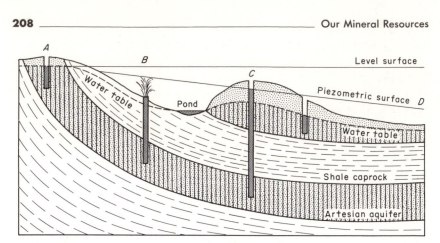

Fig. 66. Diagrammatic cross section of artesian conditions showing relationships of several wells.

produces water without the need of pumping. The well at C is also an artesian well, but requires pumping to lift the water the rest of the way to the surface. Plane $A–D$ is the result of internal pressure or head within the aquifer and is called the *piezometric surface*. It slopes downward from point A because of friction which retards the flow of water in the water-bearing stratum. Were it not for this friction, the water would rise in wells everywhere to the exact level of point A. It is apparent that wherever the piezometric surface lies above the surface of the land a free-flowing artesian well is possible. Many artesian springs issue from fissures which extend through the impermeable cap rock and into the aquifer. It is a common practice for geologists to make a contour map of the piezometric surface using information gained from many wells which have been drilled to the artesian aquifer. By comparing such a map with the surface topography one can easily locate other flowing wells. Unfortunately the piezometric surface may be greatly lowered over a period of years when too many wells producing from the aquifer lower the internal pressure.

Ground Water in Coastal Regions and Islands

Beneath the oceans all the ground water is salty except where very special conditions prevail. Under islands and in coastal areas the ground water is fresh if the land areas receive any rainfall at all. Where the fresh ground water meets the salty ground water there is some mixing, but for the most part the two stay remarkably separate. For this reason the fresh water floats upon the more dense sea water just as oil floats upon water. The fresh water under an island occurs as a lens which like an iceberg has its major part below sea level. An equation called the Herzburg formula enables us to determine the amount of fresh water in this lens below sea level. In Fig. 67 at any one well t is the elevation of the ground-water table above sea level, and h is the depth below sea level that fresh water extends. The total thickness of the fresh water lens is H, which is equal to $h + t$. It is apparent that the weight of the column of fresh water H is equal to the weight of the column of salt water h which is displaced. This is easily shown by the familiar law of floating objects.[5] If the specific gravity of fresh water is 1 and that of salt water equal to g, we can express this weight equivalence by the formula:

$$hg = 1H = h + t$$
or
$$h = t/(g - 1)$$

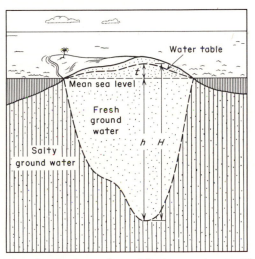

Fig. 67. Cross section showing how a lens of fresh ground water "floats" in the salty ground water under an island. (Vertical dimensions foreshortened.)

---

[5] Archimedes discovered that any mass floating in water displaces an amount of water equal to its own weight.

By substituting the average specific gravity of sea water 1.025 into the formula, we get the relation:

$$h = 40t$$

On an island where the ground-water table is 5 feet above sea level a well may be sunk to a depth of 200 feet below sea level before encountering salt water.

Effect of the Pumping Well

In a well drilled to an unconfined aquifer the water will always stand in the pipe at the level of the ground-water table when the pump is not operating. As soon as the pump starts, the water level in the casing lowers, and water is drawn from the aquifer through the well screen. Thus, a flow is started that brings water from all directions to the well. This causes a lowering of the water table which is most pronounced close to the well and is less farther out. Because of its shape this induced "dimple" in the ground-water table is called the *cone of depression* (Fig. 68). The amount that the level in the well is lowered below the ground-water table by pumping is called the *drawdown*. The *radius of influence* is determined by the farthest distance from the well that the water table is affected by the pumping.

The cone of depression gradually enlarges around a well pumping at a constant rate and may not reach its maximum size until as much as 48 hours of pumping. It has then reached a sort of a balance whereby its slope is adjusted to a flow of water toward the

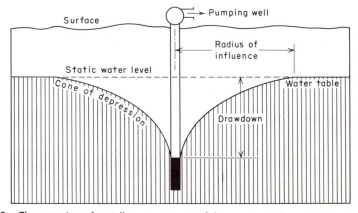

Fig. 68. The pumping of a well causes a cone of depression in the ground-water table.

well which is equal to the rate of withdrawal by the pump. An increase or a decrease in the rate of pumping will have a corresponding increase or decrease in the size of the cone of depression. When pumping ceases the cone gradually fills until the water table is restored to its original position. The rate of recovery of the water table also depends upon the permeability of the aquifer.

The permeability also affects the shape of the cone of depression. In dense materials with low permeability water enters the well screen slowly. Hence, pumping will cause a great drawdown with a radius of influence that is rather small. The steep-sided cone which results is in keeping with the principle that low-permeability soils can maintain steep slopes in the ground-water table. On the other hand, a well pumping at the same rate in a permeable aquifer would cause a cone of depression that has a small drawdown and a relatively large radius of influence. The gentle slope of such a cone is adjusted to the easy flow of water through the aquifer at a velocity just great enough to maintain the rate of pumping.

It is of great importance to know the size and shape of the cone of depression of a well pumping at its normal yield if there are to be other wells nearby. If wells are so closely spaced that their cones of depression overlap, their maximum yields will be proportionately decreased. Exact figures for the radius of influence are impossible to state, but for a well in silty sand a radius of up to 300 feet could be expected. In gravel the radius of influence may be as much as 2400 feet, and for material with intermediate permeabilities, radii may fall between these extremes. It is possible for a large well to have such a large and deep cone of depression that it could cause nearby shallow wells to go dry. Even ponds or marshes within the radius of influence may be partially drained by the pumping well. The amount of drawdown in wells producing from islands or coastal regions is of great importance. According to the Herzburg formula, every foot of drawdown of the water table causes a rise of 40 feet in the fresh water–salt water contact. In other words, the cone of depression is mirrored by a cone of salt water 40 times as steep. For this reason, such wells may produce brine if they are pumped too fast.

Quality of Water

Almost as important as the abundance of a supply of ground water is its quality. Only distilled water comes close to being pure H_2O, and all ground water contains measurable quantities of dissolved mineral matter and gases. In some cases these impurities are present in amounts great enough to make the supply of water useless for many purposes. Such contaminated ground-water sources can only be utilized after special treatments are devised.

One of the most harmful minerals that may contaminate ground water is common salt. It cannot be removed economically, but slightly brackish waters can be used for some purposes. Most live stock can tolerate water which is far too salty for human consumption, but water with only half the concentration of sea water will support no land plant or animal.

Ground water in its movement through the earth normally dissolves the more soluble constituents of the material through which it is moving. For this reason, it often contains calcium and magnesium carbonate in solution. If there is more than 200 ppm of these minerals, the water is said to be "hard." Such water is perfectly good to drink but is very wasteful of soap when used for washing. It will also precipitate calcite or "boiler scale" when it is heated, a property that makes it unfit for many industrial uses (Fig. 69). Because this kind of hardness is easy to correct by boiling, it is called _temporary hardness_. Water-softening treatments using lime water and soda are used to precipitate calcium and magnesium carbonate, leaving more soluble salts in their place that will not form scale when the water is heated. A domestic water-softening treatment utilizes a mineral substance called a _zeolite_. Hard water that is filtered through this mineral is subjected to an interchange of the metallic ion which is in solution. Calcium ion is absorbed by the zeolite which gives up two sodium ions to take its place in the water. Sodium carbonate which remains in solution will neither form boiler scale nor have a serious effect on the use of soap.

Some waters contain harmful quantities of calcium and magnesium as sulfates or chlorides or in other forms so soluble that boiling will not cause precipitation. Such waters are said to have _permanent hardness_ because the condition can be corrected only by complex and expensive treatment.

Iron and manganese salts are sometimes present in ground water. They are undesirable because of the way they stain fixtures, clog pipes, and interfere with many chemical uses for the water. Usually the addition of chlorine to the water will cause these metals to precipitate as hydroxides, after which they can be easily filtered out.

Ground waters sometimes dissolve gases which must be removed before the water can be used. Carbon dioxide in excess causes

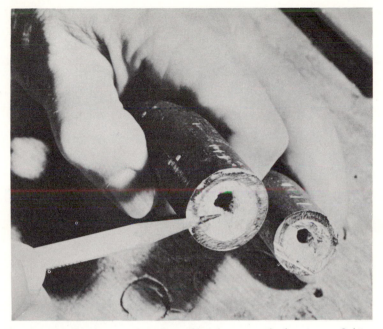

Fig. 69. Photograph showing the effect of hard water on boiler pipes. Calcium carbonate deposit has nearly closed this pipe. (Photo courtesy of Culligan Soft Water Institute.)

natural "soda water" which is often exceedingly hard. Gases of sulfur such as hydrogen sulfide may be present, particularly in regions of volcanic activity, and give the water a noxious odor. Methane gas is also common and indicates a history in which the water was in contact with organic deposits such as petroleum, coal, or decaying vegetation. All these gases are easily eliminated by a simple treatment called *aeration* whereby the water is sprayed into the air or is caused to flow over trays of rock and charcoal.

GROUND WATER OF THE HAWAIIAN ISLANDS

The Hawaiian Islands offer excellent examples of several different kinds of ground-water accumulation which are closely related to the geological conditions. The entire chain of eight major islands are but the tops of volcanic mountains that rise to a maximum height of 30,000 feet above the sea floor at the two peaks Mauna Kea and Mauna Loa (each nearly 14,000 feet above sea level). All of the islands have formed by the accumulation of thin layers of basaltic lava of a highly porous nature. Many dikes and some major faults intersect the bedding of the lava flows. The weathering of the mountainous interiors has resulted in the accumulations of aprons of sedimentary debris around the flanks of many of the islands. Interbedded in this material are layers of limestone which were ancient coral reefs and layers of fossil shell fragments. The effects of stream erosion, valley filling, wave erosion on the coasts, and geologic changes in sea level have all been factors in shaping the present topography of the islands.

The ground-water sources in the islands (Fig. 70) fall into five distinct types each requiring different methods of recovery.[6]

1. Basal ground water
   a. Unconfined with amounts based upon the Herzburg formula
   b. Artesian
2. Perched or high-level ground water
   a. Confined between dikes
   b. On old stream deposits
   c. On beds of ash or tuff

The basal ground water, where it is unconfined, is often in short supply. Because of the very high permeability of the lavas, the fresh water runs rapidly to the sea. The water table does not stand very high above sea level. Consequently, the fresh water lens does not extend very deeply below sea level. The confined or artesian sources are caused by a layer of clay at the base of the apron of sediments. This acts as a cap rock for the water in the underlying lavas, and its seaward dip creates a condition of internal pressure near the coast. Many flowing wells tap this supply.

The perched or high-level ground water is prevented from flowing freely to the sea because of layers of impermeable rock which lie below or around the porous aquifer. On Oahu a swarm of vertical dikes extend along the Koolau Range in a subparallel fashion. The dike rock is dense and impervious so that these sheetlike bodies act like walls to hold water at a high level. This is a copious supply of water now exploited by horizontal tunnels that penetrate the dikes at a low level. Old cemented stream deposits and beds of ash are in many places extensive enough and suitably impervious to act as a ground-water "floor." Lying on these beds are lenses of water with a perched water table high above that in the surrounding lavas. These water supplies are also tapped by tunnels from below the floor.

---

[6] Classification by R. G. Sohlberg tabulated from pp. 539–552 of _Ground Water,_ by C. F. Tolman. Copyright 1937 McGraw-Hill Book Co., New York.

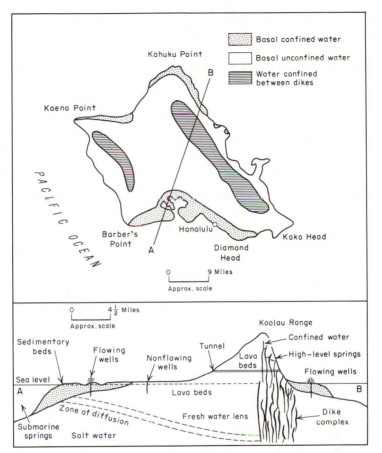

Fig. 70. Map and cross section of the island of Oahu showing various types of ground-water occurrences. Vertical scale of the cross section is greatly exaggerated. (By permission from *Ground Water,* by C. F. Tolman. Copyright 1937, McGraw-Hill Book Co.—modified by R. G. Sohlberg from Stearns and Vaksvik, *Hawaii Terr. Hydrography Bulls.* 1 and 2.)

DAKOTA SANDSTONE ARTESIAN OCCURRENCE

Flanking the Black Hills and dipping away from it in all directions is a porous unit of sandstone over 100 feet thick which has been named the Dakota formation. Lying directly above is a thick layer of impervious shale. The conditions necessary for artesian flows exist where the Dakota sandstone is tapped at a lower level in western and central South Dakota. The water entering the sandstone at the Black Hills is confined to this stratum by the overlying caprock of shale. It may percolate many hundreds of miles before it is withdrawn by artesian wells in the Great Plains where it is so badly needed. This great aquifer has been a major source of water for many years, although excessive use of the supply has greatly decreased the internal pressure. In certain areas a decline in "head" has been recorded of over 300 feet since 1900.

NEBRASKA SAND HILLS[7]

This great area in west central Nebraska consists of 20,000 square miles of grass-covered sand dunes that act like enormous sponges. Nearly all of the precipitation which falls in this area soaks into the ground and enters the great reservoir provided by the porous sand dunes and by as much as 1200 feet of Tertiary sandstone which lie below. This water slowly migrates to the southeast, and its seepage feeds the three Loup rivers and the numerous tributaries which enter the Platte. Much of the water percolates through the sandstones under the Platte valley and finally enters the surface drainage in the valley of the Republican River. From data on the numerous wells in the area, it has been estimated that the area contains a readily usable reserve of 600,-000,000 acre-feet of interstitial water. If such an amount of water were all at the surface, it would cover the area with a lake over 45 feet deep.

SELECTED REFERENCES

Barkley, R. C. *Artesian conditions in southeastern South Dakota,* South Dak. Geol. Surv. Rept. Inves. No. 71, 71 pp., 1952.

Bennison, E. W. *Ground Water, Its Development, Uses and Conservation,* Edward E. Johnson, St. Paul, 1947.

Fox, C. S. *Water; A Study of Its Properties, Its Constitution, Its Circulation on the Earth, and Its Utilization By Man,* Philosophical Library, New York, 148 pp., 1952.

McGuinness, C. L. *The water situation in the United States, with special reference to ground water,* U. S. Geol. Surv. Circ. 114, 127 pp., 1951.

Meinzer, O. E. *The occurrence of ground water in the United States,* U. S. Geol. Surv. Water-Supply Paper 489, 1923.

Meinzer, O. E. *Outline of ground-water hydrology,* U. S. Geol. Surv. Water-Supply Paper 494, 1923.

Sayre, A. N. Ground water investigations in the United States, *Econ. Geol.,* Vol. 43, pp. 547–552, 1948.

Tolman, C. F. *Ground Water,* McGraw-Hill Book Co., New York, 1937.

Wentworth, C. K. *Geology and ground water resources of the Honolulu–Pearl Harbor area, Oahu, Hawaii,* Honolulu, Bd. Water Supply, 111 pp., 1951.

---

[7] From A. L. Lugn, Nebraska's Underground Water, Part I—An Inventory, *Nebraska Farmer,* December 4, 1954.

chapter 12

coal

USES

Coal is probably still the world's leading mineral fuel, although its position is today challenged by petroleum. Coal is burned to liberate heat, and the heat may be used to generate electric power in the steam turbine or to provide mechanical power in the locomotive or ship. Coal may also be used to make combustible gases that are often mixed with petroleum gas and piped as a domestic and industrial fuel to metropolitan areas. Such gases and coal-generated electricity are merely more convenient forms in which to transport and utilize the great energy held in this black rock.

Coke is a fuel made by "baking" coal in large coke ovens that heat the coal to very high temperatures. The coal does not burn because air is kept from the ovens, but the gaseous part of the coal is baked out, leaving a hard spongy mass of pure carbon, the coke. The gas liberated in the process contains many valuable substances, such as tar, oil, ammonia, and a host of organic chemicals.

The gas is also used as a fuel. The metallurgical industry is completely dependent upon coke, for no other fuel can be used in the modern blast furnace. Unfortunately, not all coal is suitable for coking, and the limited supply that can be used for this purpose is sometimes misused for simple domestic fuel, a great waste of a diminishing natural resource.

WORLD PRODUCTION OF COAL

In 1953 all of the countries of the world mined a total of 1,964,000,000 tons of coal of all ranks. In spite of competition provided by the ever-growing petroleum production, this is more than an 8% increase over the 1950 yields. Germany and U.S.S.R. produced 400 and 320 million tons respectively. Mines in the United Kingdom yielded 228 million tons. Nearly 94 million tons were mined in Poland; and China, France, and Czechoslovakia had coal production of about 56 million tons each. Japan, India, and Belgium also had significant production. The United States led the world with over 535 million tons of bituminous and anthracite in 1956. Much of the other tonnages noted represent lignite, a coal of lower rank and heating ability. Two-thirds of the German production and one-fourth of the Russian coal was of this kind. Czechoslovakia and Poland also mine much lignite.

CLASSIFICATION OF COAL

Coal is classified into three main types or "ranks:" (1) lignite, (2) bituminous (Fig. 71), and (3) anthracite (Fig. 72). The word "rank" is well chosen, because the coal in its process of formation progresses from one rank to another. All anthracite was once bituminous and all bituminous coal was once lignite. Various intermediate stages are also recognized. An initial rank called *peat* (Fig. 73) is not considered a form of coal, yet it, too, can be used as a fuel.

Peat. This a brown porous mass of partly decomposed vegetable material. It burns readily after it is dry but does not yield much heat. All coals were once peat.

Lignite. Lignite is sometimes called brown coal because of its dark-brown color. It has a layered appearance and commonly contains many pieces of wood imbedded in a pulpy structureless mass.

Fig. 71. Bituminous coal; note bedding and blocky fracture.

Fig. 72. Anthracite coal; note shiny conchoidal fracture.

Fig. 73. Peat.

Because of its high water content it shrinks, cracks, and usually disintegrates when it is dried in air. It is the lowest rank of coal and burns with a long smoky flame, but it is used for local domestic fuels. It can also be used for the generation of gas and the production of synthetic gasoline.

Subbituminous. This coal is a rank intermediate between lignite and bituminous. It is often called black lignite because it splits parallel to the bedding planes and checks slightly when it is dried. It does not display the secondary blocky cleavage.

Cannel coal. This coal is a variety of bituminous which is dull and massive without any bedding layers. It is dense, with a conchoidal fracture, and does not soil the fingers. It burns readily but is not as good a fuel as ordinary bituminous. Because it is clean, cannel is preferred for fireplaces, and some has even been used as a decorative building stone.

Bituminous. This coal is denser than lignite and has a lustrous black color. It contains little water and does not check when it is dried. Bituminous coal is characteristically banded and breaks parallel to the beds. A secondary fracture at right angles to the beds gives the coal a blocky appearance. It burns readily with a smoky yellow flame but with greater heating power than lignite. More bituminous coal is mined than any other kind because it is plentiful, has the greatest fuel value, and it can be adapted to more special uses.

Semianthracite. The rank of this coal is intermediate between true anthracite and bituminous.

Anthracite. Anthracite is the hardest coal and has a jet-black color with a very shiny luster. Banding is less distinct than in bituminous, and the fracture is always conchoidal. It is the highest rank of coal and is an excellent fuel. It does not ignite easily but burns with a short hot flame and gives little smoke. It is widely used for domestic heating because it is clean and nearly smokeless.

COMPOSITION OF COAL

Chemically, coal is a very complex and extremely variable mixture of substances, derived almost entirely from the carbohydrates, cellulose and lignocellulose (compounds of carbon, hydrogen, and oxygen), which are the principal constituents of all plant material. There are hundreds of chemical compounds which can be extracted from coal, and a complete chemical analysis is not only very difficult but of little practical value. A simplified chemical study called a *proximate analysis* includes tests for water, volatile matter, fixed carbon, ash, and sulfur. Other properties which determine the coal's usefulness are also described.

Water is most common in the lower rank coals, reaching a maximum of about 43% in the average lignite but amounting to less than 4% in anthracite.

The *volatile matter* in coal can be driven off when it is heated to red heat. Carbon dioxide and carbon monoxide are both usually present in large amounts. A variety of chemical compounds of hydrogen and carbon called hydrocarbons are also of great importance and with hydrogen are evolved from most coals as volatile constituents. Hydrogen and the hydrocarbons have high heating value, and some bituminous coals produce more heat than some anthracite coals because of them.

Fixed carbon means carbon which is not combined with any other element. Coals with the highest rank contain the most fixed carbon. Anthracite has as much as 96% and lignite as little as 38%. Except for the few coals rich in hydrogen and hydrocarbons, the ones with the most fixed carbon make the best fuels.

The *ash* which is left after the burning of a coal is caused by the mineral impurities which are present in the coal. Clay, silt, iron oxides, and other mineral substances are common. When a coal has a high percentage of ash, it is said to be a low *grade* of coal. The grade of a coal has nothing to do with its rank, for an anthracite might have formed with a high percentage of mineral impurities whereas a lignite may have a very small ash percentage. For many uses it is very important to know at what temperature the ash will melt. If the temperature of the furnace rises above the melting point of the ash, large porous *clinkers* will form from the molten mineral. The hot clinkers form a pasty mass that can clog the grates of a furnace and make it operate poorly.

Most coals contain some *sulfur,* mainly combined with iron in the minerals pyrite and marcasite. When the coal is burned sulfur is given off as sulfur dioxide gas. Too much sulfurous gas in the smoke from a coal acts like an acid on the furnace, stack, or anything else it encounters. Therefore, for domestic uses and many industrial uses coal with only a small sulfur content is desirable. The iron from the pyrite and marcasite remains with the ash.

Another important property of a coal determined in the proximate analysis is its *calorific value.* This indicates the quantity of heat which the coal can produce. It is measured in terms of the number of British thermal units[1] (Btu) that are produced by 1 pound of the coal or the number of large calories[2] that 1 kilogram of the coal will yield. The calorific value of lignite is about 7500 Btu and that of a high rank of bituminous is over 15,000 Btu. Anthracite has a slightly lower value than bituminous.

ORIGIN OF COAL

Everyone now agrees that coal is formed almost entirely from the accumulation of plant material. However, many special conditions must prevail before there can be an accumulation. The most important condition is that the dead wood be protected from the chemical attacks of the atmosphere. In the rain forests of the world, such as the jungles of the Amazon and Congo and the great conifer forests of our Pacific Northwest, vegetation grows at a rapid rate, and much is always dying. Peat does not accumulate in such forests, however, because all dead vegetation is at once attacked by the oxygen of the air. In effect, the wood burns but at a very slow rate. Its carbon is given off as carbon dioxide, and much water is liberated as well. After a decade or so, even a large tree trunk is spongy and "dry rotted." The process is aided by the activity of termites, fungi, and bacteria.

Should the tree die and fall in a marshy area, it would begin to decay but would soon become water logged and sink below the surface of the water. Here it is protected from the atmosphere, and the process of decay soon stops. Some bacterial disintegration makes a gelatinous ooze of much of the woody material, but the little oxygen dissolved in the water is quickly used up before it can affect much of the wood. As long as the water remains stagnant and no more oxygen is introduced, the wood is protected from further attack. In this way, thick layers of partly decomposed pulpy masses of vegetation form at the bottoms of swampy lakes. This is peat.

In order for peat to accumulate in a relatively pure state, the swamp should be one in which no other sedimentary material is collecting. If streams flowing into the swamp bring in much clay,

---

[1] A _British thermal unit_ is the amount of heat necessary to raise the temperature of 1 pound of water one degree fahrenheit.

[2] A _calorie_ is the amount of heat required to raise the temperature of 1 gram of water one degree centigrade. A large calorie equals 1000 gram calories.

silt, or sand, the peat will be contaminated by these substances and eventually become a coal of low grade. These minerals make up the ash content of the coal. Stream or tidal currents in the swamp would constantly bring in fresh oxygenated water and might also flush away the woody material before it could accumulate. Flooding due to seasonal rains would have the same effect, so that a climate with an even distribution of rainfall over the year must have prevailed at the time when the coals were forming. Temperatures were probably moderate, and freezes were surely rare. The fossil plants found in coals are like some warm-weather plants which grow today in swamps in Florida, Georgia, and Louisiana. Ferns, rushes, flowering plants, and conifers are represented, but nearly all were giants compared to their modern counterparts (Fig. 74). Many had the bulbous and arched roots which are typical of plants that grow today in fresh water.

Peat deposits can never become coal until they are subjected to the pressure of overlying layers of sedimentary rocks. The seas of the past spread across the old swamps and buried the peat with layer upon layer of sandstone, shale, and limestone. Some heat

Fig. 74. Reconstruction of a forest of Pennsylvanian times. Such vegetation growing in swamps ultimately changed to the coal we are burning today. (Photo courtesy Chicago Natural History Museum.)

may have helped in the coal-making process, for the chemical changes which occurred generated heat that could not have readily escaped. The physical and chemical changes in the peat were gradual and resulted in a slow transition first to lignite and then to the higher ranks of bituminous and anthracite. One of the most important physical changes was the compaction of the porous peat beds. This caused water to be expelled, and the gelatinous organic ooze was gradually squeezed until it developed bedding-plane cleavage. The material became harder and denser until it reached the lignite stage. The pressures decreased the thickness of the beds to as little as one-twenty-fifth of the thickness of the original layers of peat. Other changes in cleavage, density, color, and luster occurred as the coal progressed through the various ranks.

The chemical changes which caused the transition of peat to coal are not at all well understood. However, the results of these changes are easily recognized. There is a gradual increase in the proportion of fixed carbon. At the same time the hydrocarbons, water, and oxygen are slowly eliminated. Complex organic reactions change some of the hydrocarbons to heavy tarry substances. The ultimate change is a crystallization of the fixed carbon to graphite when the coal is intensely metamorphosed.

The time required for these chemical and physical changes depends upon the intensity and nature of the pressure and heat which caused the changes. In general, older coals have been more deeply buried for longer times and are usually of higher rank than younger coals. Compressional forces in the earth, the kinds that fold rocks into mountains, can greatly hasten the coalification process. Anthracite can only form where such forces have been intense. Igneous intrusions which cut beds of coal have been known to locally increase its rank to natural coke.

OCCURRENCE OF COAL

Every continent, including Antarctica, has some coal (Fig. 75), but North America has nearly 40% of the estimated reserves of the world, and United States has most of this within her boundaries. In contrast, most of the South American and African countries are lacking in this rock fuel. U.S.S.R. has very great reserves, and China has abundant high-rank coal.

The major coal fields of the world are surprisingly great areas underlain by coal beds. For economic reasons, however, only part

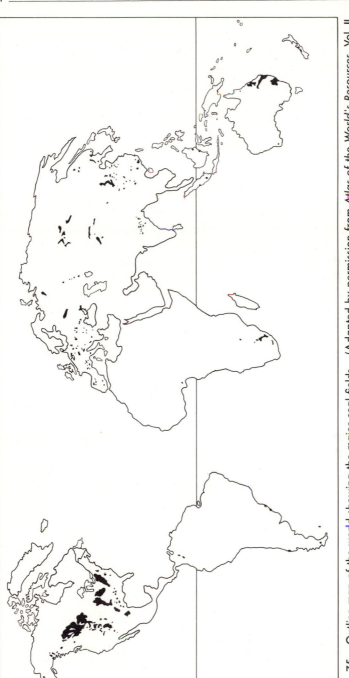

Fig. 75. Outline map of the world showing the major coal fields. (Adapted by permission from *Atlas of the World's Resources*, Vol. II, *The Mineral Resources of the World*, by William Van Royen and Oliver Bowles. Copyright, 1952, by Prentice-Hall, Inc., and published for the University of Maryland.

of this coal can be extracted profitably, so that mines are not evenly distributed throughout the coal fields. It is significant that all the important industrial countries contain or have access to abundant coal. A very close relationship exists between the locations of heavy industries and coal fields.

Coal Fields of the United States

Figure 76 shows the three main geographic divisions of coal fields which can be designated as (1) the Appalachian fields, (2) the Interior fields, and (3) the Rocky Mountain fields. These fields embrace deposits which contain coals of many ranks and ages.

The Appalachian fields. These are perhaps the most important coal deposits in the world and supply over 70% of the bituminous coal and all the anthracite used in this country. The coal beds are mostly of Pennsylvanian age, but some Mississippian and Permian coals are of local importance. All have been involved in the disturbances which formed the Appalachian Mountains, and some are now steeply dipping and complexly folded. The coals which are most crumpled lie in the eastern part of the Appalachian belt. The Appalachian coals thin towards the west where they were also less affected by the mountain-building forces.

The anthracite fields lie in eastern Pennsylvania in a region where the rocks have been intricately folded and metamorphosed. The coal is found in the remnants of synclines which are all that are left of major folds. Probably over 90% of the anthracite coal has been eroded away. Even so, these deposits have yielded over 4.5 billion tons, and nearly twice this amount still remains. The coal is of excellent quality, but mining costs are high because of the complex structures.

The Interior fields. There are four coal-producing fields which lie between the Appalachian and the Rocky Mountains. In addition, there is a vast area underlain by lignites, which are now only used for local consumption. The principal field occurs in Illinois, Indiana, and Kentucky where so much bituminous coal is produced that only the yield of the Appalachian fields exceeds it. Most of this coal is not good for making coke but is excellent for domestic and steam uses. The mines of this area enjoy an advantageous marketing situation in the many large industrial cities nearby. The western and southwestern interior fields in Iowa, Missouri, Arkansas, Oklahoma, and Texas have a good grade of bituminous coal, but much petroleum and natural gas are also produced in the

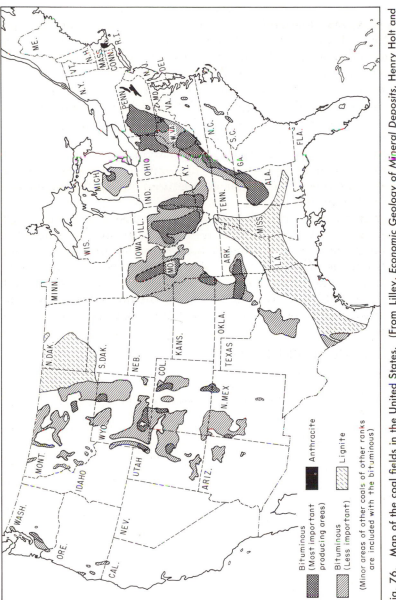

Fig. 76. Map of the coal fields in the United States. (From Lilley, *Economic Geology of Mineral Deposits*, Henry Holt and Co., 1936.)

area, and the coal industry has suffered from the competition. A small northern field in Michigan is mined for noncoking bituminous coal.

The Rocky Mountain fields. These coal deposits occur from Arizona and New Mexico northward through the mountains and intermountain basins into Alberta and British Columbia. Most of this coal is Cretaceous in age, so lignite and subbituminous ranks are most prevalent. However, good-quality bituminous coal is mined in southern Colorado and central Utah, and even excellent coking coals are produced here. High-rank bituminous and anthracite occur locally where the Cretaceous beds have suffered intense folding. Most of the deposits of the Rocky Mountain fields are mined only for local consumption or are not exploited at all because of limited markets and the strong competition offered by petroleum, which is also abundant.

There are small and unimportant coal deposits of Tertiary age in Washington, Oregon, and California. Bituminous and subbituminous occur in the Cape Lisburne area of northwest Alaska and also in the Matanuska area of southern Alaska. This coal is little worked except for local use because of transportation and marketing problems. Intense folding in the areas has caused the bituminous coal to become highly fractured, making it difficult to handle.

SELECTED REFERENCES

Campbell, M. R. _Coal fields of the United States,_ U. S. Geol. Surv. Prof. Paper 100, 1929.

Francis, Wilfrid. _Coal, Its Formation and Composition,_ Edward Arnold London, 567 pp., 1954.

Lewis, J. V. Evolution of the mineral fuels, _Econ. Geol.,_ Vol. 29, pp. 1–38, 157–202, 1934.

Moore, E. S. _Coal,_ 2nd ed., John Wiley and Sons, New York, 1940.

Stutzer, Otto. _Geology of Coal,_ University of Chicago Press, Chicago, 1940.

Wilson, Philip J., and Joseph H. Well. _Coal, Coke and Coal Chemicals,_ McGraw-Hill Book Co., New York, 494 pp., 1950.

chapter 13

petroleum

CHEMICAL NATURE OF PETROLEUM

Petroleum is a complex mixture of many chemical compounds called *hydrocarbons*. There are literally hundreds of ways in which hydrogen and carbon may combine, and most crude oils contain scores of these compounds. Three hydrocarbon series occur in important quantities: (1) the *paraffin series,* (2) the *napthene series,* and (3) the *benzene* or *aromatic series*. Crude oils from different parts of the world contain these substances in different proportions, and no two oil fields will produce exactly the same kind of oil; indeed, oil from different levels in the same field may be different.

The *paraffin series* of hydrocarbons include such familiar compounds as methane, ethane, propane, and octane. The simpler members are gases and liquids. The more complex paraffin compounds are solids. Petroleums rich in these compounds are called *paraffin-base* crudes and are the most valuable because they yield a high percentage of the best grades of lubricating oil.

The *napthene series* is made up of many compounds which are the most abundant hydrocarbons in those petroleums classed as *asphaltic-base* crudes. When distilled they leave a residue of asphalt. When the paraffin and the napthene compounds occur in nearly equal proportions, the petroleum is called a *mixed-base* crude.

The *benzene* or *aromatic compounds* are a family of light and highly combustible hydrocarbons that rarely occur in amounts greater than 10% in most petroleums. At least three other minor hydrocarbon series are known.

Complex chemical compounds containing sulfur are present in nearly all crude oil and are very harmful because of the expensive refining methods required to remove them. Oxygen and nitrogen compounds also occur in small amounts, and traces of nickel, iron, vanadium, gold, and many other metals have been detected.

REFINING OF PETROLEUM

In order for petroleum to be employed for such a great number of uses, it must be refined or separated into its many component parts, each of which has properties suited to a particular application. The most important refining process is called *fractional distillation,* a method based on the fact that the vapors of the many components of the crude oil condense at different temperatures.

The distillation is carried on in a *fractionation tower* (Fig. 77) as a continuous process. The oil is heated until most of it is a vapor which is made to rise in the tall tower. Located at intervals up to the top of the tower are plates or pans which can be kept at carefully controlled temperatures. Those at the bottom are the hottest, and the coolest are at the top of the tower. Hydrocarbon vapors that condense at high temperature form on the lower plates and the more volatile compounds condense on the cooler plates higher up. Drains remove the many fractures as they form. In this way many grades of grease, heavy lubricating oils, fuel oils, kerosene, gasoline, and naphtha may be separated from the same crude oil on successively higher plates in the fractionation tower. If the fractionation is carried on at normal pressures or under partial vacuum, the fractionation is said to be "straight-run," and the yield of gasoline is seldom greater than 25% of the crude oil.

Fig. 77. Complex distillation equipment at the Humble oil refinery in Baytown, Texas. (Photograph courtesy of Humble Oil and Refining Co.)

Over 45 years ago engineers devised another technique for the refining of oil called *cracking*. In this method up to 60% of the oil is converted into gasoline, which is the most valuable major product of the refinery. All of the hydrocarbons in petroleum are made up of characteristic groups of hydrogen and carbon atoms called *molecules*. The gasoline fraction consists of small simple molecules and the heavier oils have larger and more complex molecules. In a cracking plant the petroleum is distilled under pressure in such a way that the large molecules are "cracked" or split into the smaller gasoline groups. It has been found that certain catalysts when added to the stills greatly increase the yield of gasoline.

Another method makes use of a process called *polymerization* in which molecules that are lighter and smaller than those in gasoline are made to combine at low pressures and in the presence of suitable catalysts. The process is just the reverse of cracking and further increases the yield of gasoline. It can also be used to increase the production of high-quality lubricating oils and greases.

USES FOR PETROLEUM PRODUCTS

An indication of the way petroleum is used lies in the proportions of different substances refined from the crude oil. Gasoline now accounts for 45% of all the refinery output and is used in the millions of automobiles, trucks and other vehicles. It is interesting that before 1910 petroleum was refined in such a way that a maximum of kerosene was produced, and only about 10% became gasoline. Today 37% of the crude oil is converted to fuel oils. Many grades of this oil are produced, some suitable for diesel engines, a different grade for domestic furnaces, still a different weight for bunker fuels, etc. To provide the necessary heat to operate the refineries, approximately 5% of the petroleum is burned as light gases. Kerosene has again become a fuel of importance with its use by jet aircraft. A little over 5% of the crude oil is converted to this substance. In the heavy fractions of the crude oil many useful products are present. Lubricating oils and greases are most important (4%), but asphalt, paraffin, and petroleum coke are also produced. Some of the heavier oils and sometimes crude oil are used for road surfacing.

The chemist has been able to convert the hydrocarbons in petroleum into a host of diverse products. Explosives, synthetic rubber, perfumes, dyes, medicines, paints, saccharine, antiseptics, solvents, and thousands of organic chemicals are some of the variety of products which have their origin in this evil-smelling liquid.

WORLD PRODUCTION OF OIL

[1] There has been a steady increase in the production of this valuable fuel in response to the growing demands of our mechanized world.   In 1957 the world production came to 6,419,000,000 barrels[2] of crude oil, a 5.2% increase over the preceding year.   Oil fields in the United States yielded about 41% of the world's total with a fabulous 2617 million barrels.   In 1957 oil wells in the Middle East pumped 1287 million barrels, and the American-owned fields of Venezuela yielded 1041 million barrels.   At the same time U.S.S.R. produced an estimated 663 million barrels, Indonesia 114 million, Mexico 87 million, and the growing Canadian oil industry marketed 184 million barrels.   Borneo, Colombia, and Western Europe also had significant production (Fig. 78).

ORIGIN OF OIL

No one is yet certain just how petroleum is formed, but there has been much speculation, and theories were formulated over 100

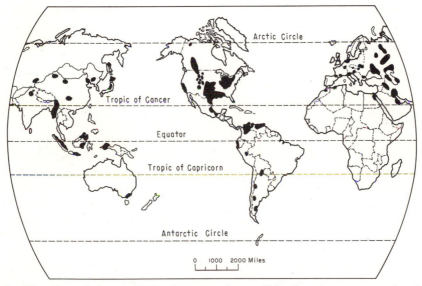

Fig. 78. Map of the world showing areas from which oil and gas are produced. (From Lallicker, *Principles of Petroleum Geology*, Appleton-Century-Crofts, 1949.)

---

[1] Statistics from *World Oil Magazine,* Gulf Publishing Co., Houston.
[2] A barrel (bbl) of petroleum is equivalent to 42 gallons.

years ago. It was once thought that coal heated with limestone deep in the earth could have formed oil compounds, but for such reactions temperatures of over 3000°F are required. Another theory stated that some lavas were the source of the hydrocarbons, and a third proposed that petroleum is just another substance that was formed at the same time as the earth was formed.

Most scientists now believe that oil is of organic origin and formed in marine muds which contain accumulations of microscopic plant and animal remains. These creatures, many of which are one-celled organisms, live even today in great numbers near the surface of the sea and provide the ultimate source of food for nearly all marine life. They are called *plankton,* and the most important are the plant forms such as *diatoms* and *algae.* There is no agreement as to the exact process whereby these organisms are converted into petroleum, but it is believed that bacteria, pressure, moderate temperatures, and great lengths of time are important.

OCCURRENCE OF PETROLEUM

Source Rocks

Organic marine muds are compacted into shales which in most oil-producing regions are considered to be the source rocks in which the petroleum originated. Rarely limestones, siltstones, and even lake sediments are source rocks. It was once thought that petroleum was generated only in black shales, but now we know that nearly any fine-grained sedimentary rock which is black, gray, or even dark brown could be a source.

Migration of Oil

Petroleum which remains in the fine-grained source rock is of very little use to man, because it does not flow out of the rock fast enough to be produced by a well. Fortunately, there is a natural migration of oil from the source rock to strata that have larger openings and are permeable enough to allow production. This movement of oil can take place over a long period of time, for we know of examples where a source bed has been deeply buried, then folded, and truncated by erosion. Beds deposited upon this surface at a later time had become saturated with oil which continued to move from the source bed even after all these events. The cause for the movement of oil into rocks with larger openings has been a subject of study and speculation by geologists and engineers for many years. In the light of all the evidence only four motivating forces are considered important: (1) compaction, (2) buoyancy effect, (3) capillary effect, and (4) water flushing.

Compaction. The compaction of oil-bearing shales caused by the weight of many hundreds or thousands of feet of overlying rocks may first squeeze much of the water out of the rock and then the oil and gas which it contains. Some muds can be compacted to less than half their original thickness, and the water and oil forced out may find their way into porous rocks above or below the shale.

Buoyancy effect. Salty connate water derived from the ancient oceans occupies the pores of the marine sedimentary rocks along with any petroleum that may be present. We all know that oil and water do not mix, and when the two occur together oil floats to the top. Thus the buoyancy of petroleum in water-saturated rocks causes the oil to move upward through rock openings to the highest point that it is able to go. Beneath the oil in nearly every field is salt water.

Capillary effect. Water has a greater surface tension than does oil. Consequently, it has a greater tendency to move into very small openings. If a shale saturated with oil is in contact with a sandstone with coarser openings filled with water, a gradual displacement of the oil by water occurs. Because of its higher surface tension, the water actually forces the oil out of the subcapillary openings of the shale. This is a slow process, and the distances through which the liquids move are measured in inches. However, once in the larger openings, the oil moves easily in response to the buoyancy force.

Water flushing. Movement of water through the rocks because of compaction or because of artesian pressures is probably common. The velocity of such subterranean flow may be as little as 1 inch per year, but even moving at such a rate, water can flush out droplets of oil and carry them great distances if given enough time. Accumulations of oil may have been flushed out and dissipated by such currents, and oil is thought to have moved in large-scale migrations because of this effect. A very slight solubility of oil in the water and an appreciable solubility of natural gas may have had more than a negligible effect.

Reservoir Rocks

Rocks that are porous and permeable enough for petroleum to be produced from them are called reservoir rocks. Sandstones and conglomerates generally meet these requirements and are the most common reservoir rocks. However, if these rocks are tightly cemented they can neither contain petroleum nor allow its flow. Limestones and dolomites may have original porosity as in coral reefs or may develop many large and connected openings by solution. These make excellent reservoir rocks which can yield copious flows of petroleum because of their high permeability. Fractured shales and jointed igneous and metamorphic rocks can produce petroleum under favorable circumstances.

OIL TRAPS

Oil which has moved from a source rock into a permeable reservoir rock tends always to migrate upwards through the rock pores because of the buoyancy effect. The upward migration may be deflected and finally stopped because of an impermeable rock barrier under which the oil may accumulate in the pores of the reservoir rock. This is called an _oil trap_. The impermeable barrier is known as a _cap rock_ and usually consists of some fine-grained or dense rock which the oil cannot penetrate. Shale, salt, gypsum, and dense limestones may act as cap rocks. Traps have been classified into two major categories: (1) _Structural traps_ occur as a result of folding, faulting, intrusion, or other events which have left reservoir strata and cap rocks in situations that resemble "upside-down" containers. (2) _Stratigraphic traps_ are the result of lateral and vertical changes in permeability of the reservoir rock caused by variability in conditions during its deposition.

STRUCTURAL TRAPS

Figure 79 shows many of the different kinds of structural traps from which oil has been produced. Anticlines and domes are probably the most important of these and are traps from which more oil has been produced than from any other kind. In such folds oil migrates up the limbs of the structure and collects at the crest below a caprock. If the fold is asymmetrical, the most oil is

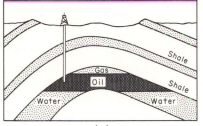
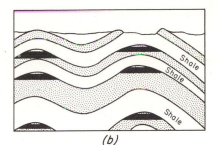

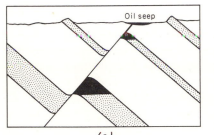
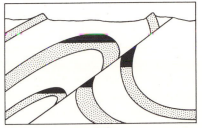
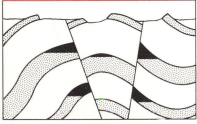

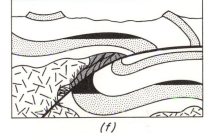
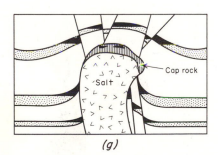
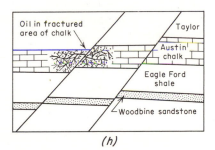

Fig. 79. Diagrammatic cross sections showing a variety of structural traps in which oil can accumulate. (a) Simple anticline or dome, (b) two anticlines, several reservoirs, (c) gravity fault, (d) thrust fault cutting anticline, (e) graben at crest of anticline, (f) overthrust, oil in fractured granite, (g) many traps around salt dome, (h) oil in fractures of brittle rock. ((h) only from Hood, *University of Texas Bull.* 5116, 1951.)

to be found on the more gently dipping limb. There is generally a level interface between the oil and the underlying water. Above the oil is usually a lens of natural gas which occupies the pores of the reservoir rock just under the crest of the fold. Unequal pressures in opposite limbs of the fold or movement of water below the oil can cause the water-oil interface to be tilted. Even in the early stages of the industry, it was recognized that such folded structures contained oil. As early as 1885 I. C. White proposed the "anticlinal theory of oil accumulation." Synclines can only yield oil when the rocks contain no water, and few instances of this are known. Monoclines are important in some areas.

Faulting of inclined strata may cause a reservoir rock to be blocked off at the fault plane by an impermeable rock. Even the fault "gouge," the ground-up rock in the fault zone, may act as a seal beneath which oil may be trapped. In the region around the Gulf of Mexico and in some other parts of the world structural oil traps have formed because of the forceful injection of salt domes into the sedimentary strata. Rocks which have been dragged up along the edge of the dome make excellent traps where they terminate against the impermeable salt. Above the salt dome beds are commonly arched up in such a way that oil can accumulate (Fig. 80).

Fig. 80. A well being drilled on a salt dome located by seismic exploration in the Gulf of Mexico 25 miles offshore from the Mississippi delta.

STRATIGRAPHIC TRAPS

Stratigraphic traps are far less obvious than structural traps, for they rarely have any indication in the surface outcrops. To locate them the geologist must make deductions based on geologic and geophysical data. Figure 81 illustrates some of the more common kinds of stratigraphic traps.

Unconformities in the rock sequence often have the right conditions for stratigraphic traps. This is particularly true where the strata below the unconformity have been folded and then truncated by erosion. If shale is deposited over the surface of erosion, an excellent seal is provided under which oil may accumulate in tilted and eroded reservoir rocks. Sometimes several unconformities occur one above the other, and each may have one or more producing beds below it.

Coral reefs formed extensively in the geologic past in much the same way as they are gradually building up in the tropical seas of today. Many of these fossil ridges of organic limestone have vertical extents of as much as 1000 feet, although as they grew they may never have extended more than a few tens of feet above the sea bottom. This means that the sea floor on which they had their base gradually sank but slowly enough so that the corals and other organisms were able to build up the top of the reef fast enough to keep it near the surface of the ocean. Accumulations of reef fragments and other sedimentary material constantly filled in around the base of the sinking reef. In these beds sloping away from the reef and in the organic limestone of the reef itself oil has been able to accumulate. The famous Leduc field in Alberta and some of the West Texas fields are of this type.

Hills or ridges on ancient sea bottoms have been gradually buried by sedimentary layers, but these strata slope away from the ridge and even thin out and dome up above it, creating traps for petroleum. A hill or ridge may be one of the many different types, such as a drowned coral reef, a rising salt dome, or an erosional hill of hard rock on a flooded terrain.

One can show that every rock unit must end or change to another kind of unit if it is traced far enough to the limits of its geographical extent. A sandstone may "wedge out" towards the old shoreline along which it was deposited. With an overlying bed of shale and a gentle dip away from the wedge-out, conditions are excellent for the accumulation of petroleum. A similar situation accounted

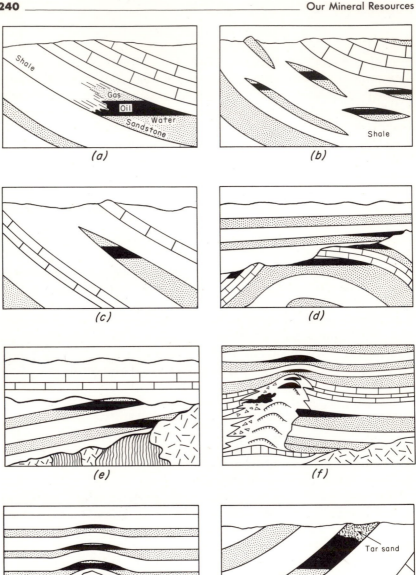

Fig. 81.   Diagrammatic cross sections showing a variety of stratigraphic traps in which oil can accumulate.   (a) Sandstone grading into shale, (b) sandstone lenses in shale, (c) sandstone bed pinching out, (d) various traps at an unconformity, (e) two unconformities and traps, (f) buried reef and related traps, (g) buried hill reflected in overlying beds, (h) seal of tar-cemented sandstone.

for the great East Texas field of 120,000 proven acres. A sandstone may grade into a shale with such a lowering of permeability
that an effective seal may occur. Shales that are source rocks may
contain lenses of sandstone that will yield petroleum.

Along many present-day shorelines there are offshore sand bars
that extend for great distances and are interrupted only where
streams enter the ocean basin. Such bars of the past when buried in
shale provide small very elongate reservoirs. Some pools of this kind
in Kansas occur in what are called the "shoestring sands" (Fig. 82).

Sometimes the oil itself provides a seal beneath which it can accumulate in a reservoir rock. If the light fractions of the petroleum evaporate, the tarry residue can effectively block further
escape of the oil.

RESOURCES RELATED TO PETROLEUM

Natural Gas

Natural gas is universally associated with petroleum. It is a
mixture of the light hydrocarbons of the same series that make up
the liquid petroleum. Most oil has gas dissolved in it which
bubbles out at the low pressures that the oil encounters at the surface. Beneath the surface excess gas not dissolved in the oil col-

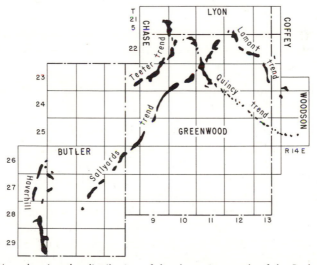

Fig. 82. Map showing the distribution of the shoestring sands of the Burbank formation in Kansas. They mark the shorelines of ancient seas. (From Bass et al., Am. Assoc.
Pet. Geol. Bull. Vol. 21, 1937.)

lects in the pores of the reservoir rock at the top part of the trap. Here it is compressed and under great pressure. The pressure which drives the oil to the surface in a flowing oil well usually is caused by the expansion of this gas. In modern engineering methods every effort is made to preserve this gas pressure so that the oil wells will have their maximum yield, and pumping may be postponed as long as possible.

In the past most natural gas was wasted because there was no way to effectively market it. Today great pipe lines carry the gas thousands of miles to cities where it is fast becoming the leading industrial and domestic fuel. In many cities large tanks store millions of cubic feet of gas for use during periods of peak demand. Recently underground storage areas have been employed that have the capacity for billions of cubic feet. These are geologic structures such as domes and anticlines that contain no oil but have a suitably porous reservoir rock beneath a cap rock "tight" enough to hold gas forced in under pressure. Great volumes of gas can be retained indefinitely in such structures, and removed by wells when needed.

Besides its use as a fuel, gas is converted in special plants into gasoline with a yield of about 1 gallon for every 1000 cubic feet. Some is used to produce _carbon black,_ required in great amounts for the manufacture of black rubber tires, printers ink, and paint pigment. The gas is burned with only a little air so that the smoky yellow flame leaves a thick coating of soot on specially cooled plates. Many oil fields do not have pipe lines for gas or produce more gas than the pipe lines can handle. This excess gas may be pumped back into the reservoir to help maintain the important pressure. Nearly all the world's supply of _helium_ is extracted from some natural gases from Texas and Utah.

Asphalt and Bitumens

If asphaltic-base crudes seep to the surface where the lighter volatile constituents can evaporate, they gradually become thicker until a heavy tarry asphalt is all that remains. Oil seeps occur where faults, joints, unconformities, and other openings provide channels through which the oil can escape. A seep may appear as little more than oil films on spring water, or it may develop into a feature as impressive as Bermudez Lake in Venezuela, a body of asphalt covering 1100 acres. Another such asphalt accumulation is at Rancho LaBrea in Los Angles, California. This asphaltic lake is famous for the wonderful fossil remains of extinct animals that were trapped in the sticky asphalt many thousands of years ago.

All deposits of asphalt do not occur at the surface. Some rocks even contain asphaltic cementing material as in the famous Athabasca asphaltic sandstones in northern Alberta which contain enormous reserves that may some day be utilized. Geologists are not sure whether such deposits were formed where oil had partially evaporated or that the asphalt was deposited in some manner at the time the sand accumulated. If evaporation is the cause for the asphaltic sandstones, one would expect to find oil down the dip of the bed beneath the impermeable asphaltic "plug." Some small oil pools have been found in this way, but too many wildcat wells drilled on the basis of this theory have been dry holes.

Asphalt and other naturally occurring bitumens are most valuable for paving material, but floor tile, roofing, pipe coatings, waterproofing, paints, and other such uses are also important.

Oil Shales

Some shales contain an appreciable amount of bituminous material which can be extracted only by distillation. Some think of this rock as being a petroleum source rock from which there has been no migration into permeable reservoir beds. These shales do not feel or appear oily and range in color from black to reddish brown. When distilled they may yield as little as 4 gallons per ton of rock and as much as 150 gallons. The United States' deposits in Colorado and Wyoming produce between 10 and 68 gallons per ton. The richest deposit is in Estonia where the oil shale yields between 70 and 80 gallons per ton. In America oil produced in this manner cannot compete commercially with petroleum, but we have enough of this rock to make a comfortable reserve should we ever need it. It is actively mined in Scotland and in other parts of the world where normal petroleum does not occur.

SELECTED REFERENCES

Clark, K. A. Athabasca bituminous sands, *Fuel,* Vol. 30, No. 3, pp. 49–53, 1951.

Heiland, C. A. *Geophysical Exploration,* Prentice-Hall, Englewood Cliffs, 1940.

Lalicker, C. G. *Principles of Petroleum Geology,* Appleton-Century-Crofts, New York, 1949.

Landes, K. K. *Petroleum Geology,* John Wiley and Sons, New York, 1951.

Leven, D. D. *Done in Oil,* Ranger Press, New York, 1941.

Levorsen, A. I. *Geology of Petroleum,* W. H. Freeman and Co., San Francisco, 1954.

Russell, W. L. *Principles of Petroleum Geology,* McGraw-Hill Book Co., New York, 1951.

Smith, P. V., Jr. Studies on origin of petroleum; Occurrence of hydrocarbons in recent sediments, *Am. Assoc. Pet. Geol. Bull.,* Vol. 38, No. 3, pp. 377–404, 1954.

VerWiebe, W. A. *Oil Fields of North America,* Edwards Bros., Ann Arbor, 1949.

rock and mineral building materials

Most mineral building materials are made from common rocks which are easily obtained and are relatively inexpensive. Stone quarries, sand and gravel pits, and clay pits are common sights everywhere in this land of ours. With bulk products such as these it is common practice for the engineer to use local resources rather than pay the cost of shipping what may even be superior materials from distant sources. Therefore, most bricks, tiles, quarry stone, gravel, sand, and portland cement are obtained from the deposits or manufacturing plants nearest to the place where they are being used. It is a valuable experience for the student to visit some of these plants and quarries to see how such materials are produced.

In this chapter many of the important nonmetallic mineral building materials are considered. For most of these, little need be said about the geology, but a brief discussion of the methods of utilization and processing is given.

244

Building Stone

Uses for Building Stone

Much stone is cut so that each individual block fits certain prescribed dimensions and has a specific place in the structure. This is called *dimension stone,* and it is usually processed at a plant near the quarries. Here stones for a specific project are cut according to plans sent by the architect. *Ashlar* is stone that is cut so that one or two dimensions are exact, but the mason on the job must cut and fit the stone further as it is used. *Rubble* and *flagstone* come from the quarry with a naturally smooth surface that is generally a bedding plane. It is not cut at all but is used as natural pieces for facing on smaller structures. Some rocks, such as slate, are valuable for roofing because of the flat thin slabs into which the rock breaks. The following kinds of rock are used in many ways as building stone.

Granite. Granite is the most common igneous rock used for building stone. Diorites, monzonites, gabbros, and other coarse igneous rocks are also used, but they are commonly referred to in the trade by such "popular" names as "gray granite," "red granite," and even "black granite." Apparently the word "granite" has an appeal as something with "strength" and "everlasting beauty" not implied by such a rock name as "gabbro." The following discussion applies to all such rocks.

Most granite used in the United States is produced in the Appalachian Mountains from Maine to Georgia. In the Midwest, Minnesota, Wisconsin, and South Dakota have important quarries. Although adequate resources are available, little granite is quarried in the western states.

Granite is probably the hardest and densest of all the building stones. It is resistant to wear and to the attacks of the atmosphere, even in smoky industrial areas. It can be found in a variety of colors and textures and is commonly given a lustrous polish. Most granite is quarried for dimension stone and is used as stone facing, columns, decorative trim, and stair cases, mostly in public buildings. It has a great appeal as a monument stone because of its beauty and durability. The cost of this rock is high because of the difficulty in cutting, shaping, and polishing such a hard material (Fig. 83).

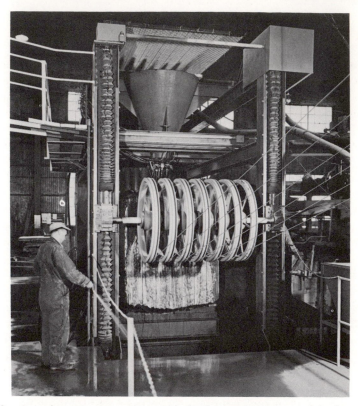

Fig. 83. A modern wire saw cutting a block of granite into seven slabs from which monuments are made. (Photo courtesy Rock of Ages Corp., Barre, Vermont.)

Limestone. Limestone is probably the most widely used building stone. It is not as costly as granite because it can be handled so much more easily. Limestone used for building is soft, usually dense, fine grained and can easily be sawed, shaped, turned, and polished. It occurs in every shade of color from nearly pure white to black, but buff and gray are most common. Because of its softness it is subject to wear when used in stairs, floors, and other such exposed places. Water can dissolve and pit the stone under some circumstances.

Most limestone is quarried for dimension stone, but much ashlar and rubble are used for homes, because a mason can easily saw and fit the stone on the job. Over 80% of the limestone sold for dimension stone in this country is from the great quarries in the Bedford-Bloomington districts of southern Indiana (Fig. 84). This stone is in great demand because of its remarkably even texture and its attractive gray or buff colors.

Sandstone. Sandstone is not used as much as it once was for building purposes. There are fads or styles in building stones, and the day of the popular "brownstone" house of the 1890's marked the last period when sandstone was in widespread use. There is still some demand for this stone as ashlar facing, for curbing, flagstone, grindstones, and linings for metallurgical furnaces.

Marble. Marbles may be classed into three major groups. The most important of these is the true metamorphic limestone. Limestones which were involved in severe mountain-building forces recrystallized as they became highly contorted by folding. Bedding became indistinct, and the rock assumed a sugary and dense texture. A second classification is the *serpentine marble,* called "verde antique" because of its green color. It contains little calcite yet takes a beautiful polish and has an ornate streaked appearance. It forms from the metamorphism of dark ultrabasic igneous rocks. The third classification of marble is *onyx marble.* This is not a metamorphic rock at all, but is formed as a layer-by-layer deposition from cold spring waters. This rock is strikingly banded, takes a high polish, and has an attractive transluscent luster.

Fig. 84. Channeling machines making long cut in quarry of Indiana oölitic limestone. (Photo courtesy Indiana Limestone Institute.)

Most marble is used for dimension stone, both as exterior and interior finish. It occurs in a variety of colors and patterns which are desirable for trim stone. Much marble is employed for monuments, and it has been always a favorite medium for sculptors.

Since true marbles are metamorphic rocks, the major production comes from mountainous regions. In America, the Appalachian mountain belt yields the best stone, particularly in Vermont, Alabama, Tennessee, and Georgia. Maryland produces an excellent quality verde antique. Arizona has a small onyx marble production.

Slate. Slate is a fine-grained metamorphic rock with pronounced, relatively smooth and flat cleavage surfaces. It is found in mountainous regions where shales have been intensely folded and metamorphosed. Most slate is used for roofing, but stair treads, wainscoting, trim, blackboards, and laboratory tables are also made from it. Slates may be red, gray, black, or green, but these colors can alter or bleach upon exposure. Roofing slates should have permanent colors or must at least alter evenly to a pleasing shade.

The best slate quarries are in the Appalachians in favorable localities from Maine to Georgia.

Crushed Stone

It has been only in the last 200 years or so that crushed stone has been used for building purposes in any significant amounts. In 1858 Eli Whitney Blake invented what is supposed to be the first stone crusher. Soon after this, the use of crushed rock in macadam roads and for railroad ballast increased greatly. Now such material is quarried, crushed, and screened in nearly every part of our land. Crushed stone is used for concrete aggregate, road metal, railroad ballast, filter stone, roofing granules, and as agricultural limestone.

Limestones are very plentiful, easy to crush, and have many uses as crushed rock. Fine-grained igneous rocks are excellent for this purpose. Granites, sandstones, slates, marbles, quartzites, and, in fact, nearly every rock type may be used.

Rock for Structural Ceramic Products

Even as early as the ancient Babylonian culture of Mesopotamia, technology was well advanced for making bricks out of clay. Our present methods for the manufacture of bricks differ little from those used 3000 years ago except for the use of efficient machinery. Bricks have always been important building materials, especially in regions where suitable building stones are not available. Today bricks are second only to lumber for use in modern construction. Also, many other ceramic products have been developed to satisfy special structural needs.

Kinds of Ceramic Materials

Any rock material used for ceramic purposes must satisfy at least these two conditions: (1) It must be plastic enough when wet to be molded into a desired shape. (2) It must become hard and resistant to water after it has been fired. Many other properties of a material must be considered, however, when special uses are made of it.

Clay is the most common material used for ceramic purposes, and fortunately it occurs almost everywhere. Deep weathering of impure limestone, igneous rock, glacial till, or other suitable rocks produces clays that may be valuable ceramic material. Mud deposits in glacial lakes, swamps, and in the backwaters of rivers are possible sources. Marine shales and even some wind-deposited dust, called loess, are also used. True clays are always at the surface and are the most recent sedimentary rocks. Shales can also be used for ceramic purposes but may require grinding if they are well indurated. Some shales are soft enough, however, to be treated as surface clays.

Ceramic Products

There are many different products that are made from clays and shales, and each one requires a clay with certain properties. Some products require very special clays, whereas for others nearly any ceramic material will do. Among the most common structural ceramic products are bricks, terra cotta, hollow-brick conduit, fire-proofing, sewer pipe, fire brick, and even a light weight porous aggregate for concrete.

PORTLAND CEMENT

History of cements. Over 2000 years ago Roman architects were using a kind of cement so excellent that many of the structures built at this time are still standing today. This was called *pozzuolana cement* after the town of Pozzuoli, Italy, where the necessary rock material was quarried. A partially weathered volcanic ash was finely pulverized and mixed with quick lime. When water was added to this mixture, a reaction took place between the lime and the natural glass of the ash that caused the mass to "set up" hard. Pozzuolana cements are still used today in small amounts.

In 1756 an English engineer, John Smeaton, discovered that certain impure limestones containing clay and silica calcine[1] to a product that would set even under water. This was called *natural cement* or *hydraulic cement*. It soon became the most widely used cement.

True *portland cement* was invented in 1824 by Joseph Aspdin, a mason and bricklayer of Leeds, England. He called the material "portland" because the cement so much resembles the famous Portland stone, a superior limestone widely used for building on the British Isles. This was a spendid name for a cement that was far better than the natural cements of the day. It soon supplanted nearly all other cements and is still the type most widely used today. Portland cement has been greatly improved, and several specialized types are now produced. In the production of modern portland cement the chemical proportions of lime, alumina, silica, and iron oxide must be carefully controlled and the pulverized rock is not merely calcined but is heated until it is partly melted.

Use of cement. Cement is rarely used by itself but is commonly mixed with sand, gravel, or other aggregate to make concrete. We are all familiar with the uses of concrete for building blocks, sidewalks, paved highways and airfields, steel and concrete buildings, dams, and many other purposes. Cement is also mixed with asbestos to make a product widely used for siding and roofing and for

---

[1] Calcine means to heat limestone until all the carbon dioxide has been driven off. Temperatures above 900°C are required to calcine pure calcite limestone.

construction where resistance to heat is important. Some masonry paints contain portland cement.

Rock used for cement. Nearly any material or mixture can now be used in the manufacture of portland cement so long as it supplies the proper percentages of lime, alumina, silica, and iron oxide. The following table shows the variety of rocks and other materials which are mixed to make this cement.

MATERIALS USED IN CEMENT INDUSTRY

Source of Lime	Source of Alumina	Source of Silica	Source of Iron Oxide
cement rock	clay	clay	limestone
limestone	shale	shale	clay
marl	coal ash	coal ash	shale
fossil shells	slag	slag	slag
recent shells	igneous rocks	sand	pyrite cinders
alkali waste		sandstone	mill scale
blast-furnace slag			

Table from *Industrial Minerals and Rocks,* Chapt. 8, A.I.M.E., 1949.

Manufacture of portland cement. The process of making cement is rather simple and can be summarized in three principal procedures: (1) the initial grinding and blending of the raw materials, (2) the burning of these ingredients to form the clinker, and (3) the final grinding of the clinker.

As the rock comes from the quarry it is fed into a jaw crusher that breaks the large pieces to pebble size that can be fed with water into ball or rod mills. These mills are nothing more than huge drums lined with steel and partly filled with steel balls or rods. As the drums rotate, the tumbling balls beat the rock fragments to a fine mud. The mud, called a *slurry,* is blended in a huge vat until the chemical proportions are correct.

The burning operation is done in a *rotary kiln* (Fig. 85), a furnace built in the form of a great tube as long as 350 feet and 10 feet in diameter. It sits at a gentle incline and is made to rotate several times a minute. The kiln is lined with fire brick and attached on the inside are lengths of chain. These drag on the lining to prevent the charge from caking on the walls. The slurry enters the upper end of the kiln where the temperature may be about 700°F. It instantly dries and slides gradually down toward the lower end where the fuel is burning and the temperature is as

Fig. 85. Large rotary kiln used in the manufacture of portland cement. This one is 10 feet in diameter and 325 feet long. (Photo by W. S. Craig, courtesy Ash Grove Lime and Portland Cement Co.)

high as 2700°F. At such a heat new compounds form, and the ground rock partially melts and rolls into small dark pellets called *clinkers*. The clinkers fall out of the kiln onto special cooling grates, after which they are sent to the final grinding mill.

The final grinding of the clinkers is exactly like the preliminary grinding of the rock except that it is done dry. Ball mills reduce the clinker to a flourlike consistency required for the portland cement. During the grinding, gypsum is added to make up about 3% of the final product. This has the property of making the cement set more slowly. The finished cement is sacked in 94-pound bags[2] or shipped in bulk by rail or ship.

There are many variations to the process described above. Some of the newest plants are using a flotation process to separate desired ingredients from what would ordinarily be undesirable raw materials.

---

[2] This amount of cement has a volume of 1 cubic foot.

GLASS

No one knows when or how man first learned to make glass, but articles of glass were found in Egyptian tombs that date back to the 4th millenium B.C. Glass jewelry, containers, objects of art, and even window glass were used in very ancient times. For thousands of years craftsmen and artists have used glass to create many beautiful objects, including such variety as optical lenses, stained glass windows, chandeliers, and wine goblets. Today this versatile substance is applied in thousands of practical and artistic ways.

Uses of glass as a structural material. In past years windows were about the only application of glass as a building material. Present-day builders are utilizing glass in new ways. Blocks made of glass are now in common usage to make walls that let in light yet provide excellent insulation to heat, cold, and moisture. The surfaces of the blocks are usually rippled or fluted to insure privacy. Another use of glass in modern building is for insulation. The glass is blown or "spun" into what is known as "rock wool" or glass-fiber insulation. Exceedingly fine filaments of glass when matted together form a light fireproof feltlike mass that is excellent insulation for the walls and ceilings of houses. Glass for this purpose is made from the cheapest rock materials available often with many impurities that cannot be tolerated when clear glass is made. Architects of modern buildings are making use of greater wall areas of windows, glass partitions, and even glass doors. "Picture windows," large expanses of plate glass in two layers, are now used extensively.

Composition of glass. Glass is made from molten mineral material that is allowed to cool so rapidly that crysals do not form. It is hard and brittle, generally transparent, and breaks with a conchoidal or shell-like fracture. Chemically, glass is mainly silica which is obtained in the form of clean quartz sand. So that this might melt easily, some kind of flux is added. Soda, lime, magnesia, potash, and borax are commonly used for this purpose. Lead oxide is added where the glass is to be used for optical purposes, for it gives to the glass the ability to bend light rays strongly. There are a great many metallic oxides that act as pigments in glass. Green and browns are caused by iron oxide, which gives a distinct color when present in only a few tenths of a per cent. Great care must be exercised to use sand and other chemicals that contain little or no iron if a colorless glass is desired. Other metallic oxides impart a variety of colors to glass. Some of these are:

copper (blues and reds), cobalt (blue), nickel and manganese (purple or brown), uranium (fluorescent green or yellow), and titanium (yellow to brown). Special properties such as resistance to heat or to shock, inertness in the presence of acids, unusual hardness, or special optical properties are obtained when different substances or proportions are used. There are thousands of different glasses, each suited for a certain use.

Glass sands. Sand that is to be used for glass making should be of very high purity. A silica content of over 95% is necessary, but the nature of the impurities is very important. The effect of iron oxide has been mentioned, and other pigmenting elements are equally undesirable. Some alumina, lime, or magnesia, in the sand are desirable impurities as these substances are added in the glass furnace. The sand should be rather fine and well sorted.

The Oriskany sandstone in Pennsylvania and West Virgina is very pure and provides most of the sand for the great glass industry in the eastern states. In Illinois, Minnesota, and Missouri the St. Peter sandstone is an excellent source of sand. Both of these rock units are so poorly cemented that the sand is easily mined. A less pure sand of Tertiary age in New Jersey is used for bottle and window glass.

Gypsum

This common mineral was known and used in very ancient times. Theophrastus applied the name not only to the mineral but also to the calcined plaster which was even then made from gypsum. It is a soft mineral, easily scratched by the finger nail. The color is generally white, but shades of gray, red, pink, or brown are not uncommon. Chemically, it is a hydrated calcium sulfate designated by the formula $CaSO_4 \cdot 2H_2O$.

Uses of Gypsum

Gypsum rock is not used as a building stone because of its solubility in water. However, for centuries alabaster, a transluscent sugary gypsum rock, has been employed for interior ornamental work. It is also widely used for such items as lamp bases, bowls, book ends, and novelties.

Great quantities of gypsum are added to portland cement as a retarder. Many thousands of tons of gypsum are used annually for fertilizer. Growers of legumes such as peanuts, alfalfa, beans, and peas find that ground gypsum provides the necessary sulfur and seems to help the plant assimilate potash from the soil. Finely ground white gypsum is called "terra alba" and is used as a filler in the manufacture of paper, in paints, and even as a nutrient for the growing of yeast.

As a building material *calcined* gypsum is of great importance. In the calcination process the gypsum is heated up to 350°F. This causes three-fourths of the water to be driven off, leaving a substance called *plaster of Paris*[3] with the composition $2CaSO_4 \cdot H_2O$. When finely ground plaster of Paris is mixed with water, it changes again to gypsum, forming a mass of tiny interlocking needlelike crystals that cause the so-called set. Plaster of Paris is the main ingredient of building plasters where it may be mixed with sand or used by itself for the smooth finish layer. Special substances may be added to retard or hasten the setting time. Builders use much prefabricated lath and wallboard made by great machines that spread the plaster on paper backing or sandwich it between two layers of paper. These products may be sawed and nailed and, when properly applied, equal in nearly every respect a custom job of plastering. Specially prepared pure white plaster of Paris is used for molding, statuary, orthopedic and dental plasters, and pottery.

Geologic Occurrence of Gypsum

This common mineral is present in the United States in rocks that range in age from Silurian through Tertiary. Some of the principal deposits are from the Camillus shale beds of Silurian age that are extensively worked in New York along an east-west belt just south of Lake Ontario. The gypsum occurs in lenses that parallel the shale beds and extend for several miles before thinning out. Most of the commercial lenses are between 5 and 6 feet thick and occur near the top of the shale. Mining is almost entirely underground. Extensive deposits in Michigan occur in Silurian rocks, but many workable beds are also present in Mississippian formations. Texas and Iowa are now leading states in the production of gypsum from deposits that are considered to be Permian in age. In California and Nevada, gypsum is mined from the beds of old salt lakes and also from some of the marine Tertiary formations.

---

[3] The famous gypsum quarries of Montmartre, in the north part of Paris produce a high-quality product that became known as "plaster of Paris." The name is now applied to all calcined gypsum.

Origin of the Gypsum

Sea water contains 3.5% salt by weight which consists of many different mineral substances. Gypsum, one of the least soluble of these, precipitates after 37% of the water is evaporated. Common salt (NaCl), the most plentiful mineral, is more soluble and does not precipitate until over 93% of the sea water is evaporated. For this reason beds of pure gypsum can form separately from layers of rock salt. Deposits of both minerals occur as a result of the evaporation of sea water.

SELECTED REFERENCES

BUILDING STONE

Bowles, Oliver. Economics of Crushed-Stone Production, *U. S. Bur. Mines Econ. Paper* 12, pp. 54–62, 1931.

Bowles, Oliver. *The Stone Industries,* 2nd ed., McGraw-Hill Book Co., New York, 1939.

Bowles, Oliver. *Memorial Stone,* U. S. Bur. Mines Inf. Circ. No. 7720, 6 pp., 1955.

Loughlin, G. F. Indiana oolitic limestone; Relations of its natural features to its commercial grading. Contributions to Economic Geology, Pt. I, *U. S. Geol. Surv. Bull.* 811, pp. 111–202, 1930.

CLAY AND OTHER CERAMIC MATERIALS

Parmelee, C. A. *Clays and Other Ceramic Minerals,* Edwards Bros., Ann Arbor, 1937.

Problems of clay and laterite genesis, Am. Inst. Min. and Met. Eng., New York, 244 pp., 1952.

Ries, H. *Clays: Their Origin, Properties, and Uses,* 3rd ed., John Wiley and Sons, New York, 1927.

Tyler, P. M. *Clay,* U. S. Bur. Mines Inf. Circ. 6155, 1935.

Wilson, H. *Ceramics, Clay Technology,* McGraw-Hill Book Co., New York, 1927.

PORTLAND CEMENT, AND GYPSUM

Bowles, Oliver, and H. H. Hughes. The Story of Cement, *Tr. Can. Inst. Min. and Met.,* Vol. 36, pp. 525–536, 1933.

Bowles, Oliver, and D. M. Banks. *Lime,* U. S. Bur. Mines Inf. Circ. 6884, 1936.

Eckel, E. C. *Cements, Limes and Plasters,* 3rd ed., John Wiley and Sons, New York, 1928.

Mielenz, R. C., K. T. Greene, and N. C. Schieltz. Natural pozzolans for concrete, *Econ. Geol.,* Vol. 46, pp. 311–328, 1951.

Moyer, F. T. *Gypsum and anhydrite,* U. S. Bur. Mines. Inf. Circ. 7049, 1939.

Stone, R. W., *et al.* Gypsum deposits of the United States, *U. S. Geol. Surv. Bull.* 697, 1920.

GLASS

Fettke, C. R. American glass sands, *A.I.M.E. Tr.,* Vol. 73, pp. 398–423, 1927.

minerals for chemical uses

Coal, petroleum, plants, and animals supply us with all of the great variety of organic chemicals, many of which have been mentioned in preceding chapters. The inorganic chemicals which man needs are found in the earth, the sea, and the air. Their utilization is a testimonial to man's knowledge of the chemistry and geology of the world about him. There is such a variety of substances of this kind that only a few of them can be discussed in this chapter.

Salt

This mineral, technically called *halite,* is easily recognized by its solubility in water, its distinctive taste, and the cubic nature of its crystals and cleavage (Fig. 86). Halite is generally colorless but, because of impurities, may appear white, yellow, blue, brown, gray, or even red. Chemically it is sodium chloride, NaCl.

Fig. 86.   Halite.

Salt has always been vital to man as an essential part of his diet. It was so important to the early people that its use became symbolic in their religions and its production was by government monopolies.   It is still used as a basis of taxation in some oriental countries.   Salt was carried as an important item of trade into the interiors of all the continents.   Officers and men of the Roman army were given a ration of salt, called a *salarium,* which was part of their pay.   This is thought to be the origin of such expressions as "worth his salt" and in fact the word "salary."

Uses for Salt

The use of salt in foods is the result of a compelling human need, and each of us consumes about 12 pounds of it every year.   The salt used for human consumption, however, is minor compared to the enormous quantities that are directed to industrial applications.   Salt is a raw material used in the manufacture of such common chemicals as soda, sodium bicarbonate, caustic soda, sal soda, and hydrochloric acid.   Salt and chemicals derived from salt are used in the manufacture of a variety of products such as dyes, paper, cotton thread, lacquer, and cements.   It is used in the tanning of leather, as a wood preservative, in water purification, in road surfacing, and as a flux in the refining of ores.   Salt is needed by the ceramic industry as a flux and for glazes.   We are familiar with its use in the home ice-cream freezer, but many industrial refrigeration systems also contain salt brines.   Salt is valuable in the home as a food preservative, medicine, and cleanser.   On the farm salt is utilized in cattle feed, hay preservatives, weed killers, and for many other purposes.

Production of Salt

Salt can be found in some form in nearly every country. There are many kinds of deposits, and the world's resources are almost unlimited. A steady increase in production has been recorded over the years. In 1956 a record world's total of 70,700,000 metric tons of salt was marketed from all sources. United States led the other countries with over 23,000,000 tons, most of which came from Michigan, New York, Ohio, Texas, and Louisiana. Russia produced 7,200,000 tons, and deposits in China yielded a total of about 6,600,000 tons. West Germany, India, and France each produced over 3,000,000 tons.

Geologic Occurrence of Salt

Salt is recovered from four different kinds of sources: (1) bedded salt deposits, (2) sea water and ground-water brines, (3) saline lakes and playas, and (4) salt domes. In the United States all four sources are worked commercially.

BEDDED SALT DEPOSITS

Beds of salt are very common, particularly in rocks of Silurian and Permian age. They generally occur with layers of gypsum, anhydrite, and in rare instances potash minerals, all derived from the evaporation of ancient sea water. Extensive areas are underlain by salt. An area including most of Michigan, northern Ohio, and western New York marks the extent of one thick sequence of Silurian salt. Near the center of the Michigan basin there is an aggregate salt thickness of at least 800 feet, and in central New York seven beds of salt total 318 feet. The great Permian salt beds in south-central United States are the largest known deposits of salt in the world. The area includes parts of New Mexico, Texas, Oklahoma, and Kansas and is roughly 100,000 square miles in extent. Under this vast area the salt is said to average 200 feet in thickness in 30 or more separate beds. Other countries have similar deposits.

One cannot explain such thicknesses of bedded salt by the simple evaporation of water from a cut-off arm of the sea. A few calcula-

tions using the present salinity of sea water show that the evaporation of nearly 10,000 feet of sea water would deposit only 100 feet
of salt. In wide shallow evaporative basins such thicknesses could
be deposited only in the deeper parts into which the concentrated
brine would drain, but thick beds of salt that are many thousands
of square miles in area must accumulate in a different fashion. As
no one believes that basins over 10,000 feet deep ever were present
on the continents, particularly as isolated remnants of the oceans,
another explanation is suggested for thick salt beds.

A geologist by the name of Ochsenius reported that the Caspian
Sea offers an example of how salt can accumulate to a great thickness in a rather shallow basin. On the eastern side of this body of
water is the Gulf of Karabugaz, 95 miles long, 80 miles wide, and
not over 45 feet deep. At its restricted mouth is a sand bar that
allows only surface water to enter the gulf. The water in the gulf
is over five times saltier than sea water, yet the salinity of the
Caspian is only a little more than one-third that of ocean water.
The reason for this is apparent when one examines the map (Fig.
87) of the region. The mighty Volga and many lesser streams of
fresh water flow into the Caspian, but none drain into the Gulf of
Karabugaz. The climate is dry and hot, particularly in the east
and southeast. The rate of evaporation is very high, in fact, so

Fig. 87. Location map showing the region of the Caspian Sea.

great that there is a constant current of water that flows over the bar to replace that lost by evaporation from the gulf. Ochsenius calculated that this current from the Caspian brings 350,000 tons of dissolved salt daily into the gulf. As the surface water becomes more salty through evaporation, it also becomes more dense and tends to settle to the bottom. If it were not for the bar at the mouth of the bay, the concentrated brine would flow as an underwater current back into the Caspian, and no accumulation of salt could occur. At the present time gypsum is depositing in a fan-shaped area starting at the mouth of the gulf. In the winter time sodium sulfate deposits along the eastern shores and on the shallow bottom. The salt concentration is not yet high enough for halite to precipitate.

Ochsenius, in his so-called *bar theory,* suggests that such a basin, restricted by a bar and in an arid region would gradually fill with salts (Fig. 88). However, it has been shown that wherever the surface of the earth is loaded by the accumulation of sediments the basin of deposition gradually settles under the load. If settling of the basin floor just equalled the rate of salt deposition, the basin might always be shallow, yet the thickness of salt reach hundreds of feet. This theory provides a logical explanation for the thick beds of salt included in the Silurian and Permian rocks.

SEA WATER AND GROUND-WATER BRINES

The oceans of the world are an endless reserve of salt. The 3.5% of salt in the average sea water may be extracted merely by evaporation of the water. However, only about 77% of the salt is halite, the rest being mainly sulfates and chlorides of magnesium and calcium. Some bromine, iodine, and potassium are also present in small amounts. Commercial producers of salt from sea water and brines make every effort to recover pure halite, and they also find it profitable to recover the rarer and more valuable compounds.

Many wells produce brines that are used as sources of salt. The brine may be ground water that has penetrated salt-rich sedimentary rocks, or it may be *connate* water trapped in the rock from some ancient sea. Much of the salt production from Michigan and Ohio is from natural brines.

Artificial brines are made by pumping water through wells into salt-bearing strata. When the water comes to the surface again in a withdrawal well, it is saturated with salt. Where conditions are favorable, this is a cheap, but effective method of "mining" salt.

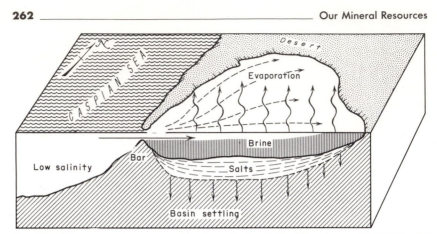

Fig. 88. Diagrammatic cross section through the Gulf of Karabugaz illustrating the "bar theory" for thick salt accumulations in a shallow evaporating basin.

SALINE LAKES AND PLAYAS

In arid regions of the world there are depressions into which many streams may flow but from which there are no drainage outlets. The "fresh" water streams carry small percentages of dissolved mineral matter in addition to mud and silt. Because water can only leave the basins by evaporation, the salts accumulate along with the sediments. This situation could only exist in an arid region where the rate of evaporation from a basin equals or exceeds the rate of inflow from the streams. In a humid climate the basin would fill with water and overflow at some point. This would create a discharge channel, and the valley would soon become part of a normal drainage system.

These arid basins may contain permanent lakes which become always more salty as the stream waters evaporate. Such lakes occur only where there is a balance between evaporation and the rate of recharge by the streams. A more common situation is for a saline lake to be only a temporary feature, drying up or partially filling the basin according to the seasons. The flat salty basin floor occupied by such an ephemeral lake is called a *playa*. Searles Lake, California is a playa, and the Great Salt Lake, Utah, and the Dead Sea in Jordan are examples of saline lakes.

The salts in these deposits are not all of the same kind nor do they occur in the same proportions as the salts in sea water. They result from the weathering of rocks in the limited areas drained by the streams. Many saline lakes are worked for their salt, using the solar method of evaporation. Salts in dry playas may be mined or quarried.

SALT DOMES

Salt domes are known in Germany, Iran, Roumania, and Spain, but those along the Gulf Coast of the United States (see Fig. 91, page 279) are most famous because of the rich occurrences of oil and sulfur that are associated with many of them. The domes are commonly 1 to 2 miles in diameter and extend downward in a pluglike fashion to great depths. No drill has ever reached the bottom of a salt dome, but geophysical studies suggest depths of between 17,000 and 20,000 feet. The typical salt plug has apparently forced its way through the surrounding sedimentary rocks, causing them to bend upwards against the salt and dome up above it. There is commonly a "caprock" above the salt that consists of anhydrite, limestone, gypsum, and sometimes sulfur. It may be up to 600 feet thick. Most caprocks thin out at the edge of the dome, but there are some that extend down the sides covering the dome like a great thimble. A few salt domes are exposed at the surface, but most are still buried by hundreds of feet of rock.

The origin of salt domes has been a subject of speculation and controversy for many years. The most recent and widely accepted views are that salt in a deeply buried bedded deposit responds to pressure like a plastic material. It is lighter than the surrounding sediments so that it has a tendency to flow upwards when buried by other sedimentary material. As the salt pushes upward a surrounding basin develops in the rocks around the dome. The thinning of beds over the dome and the nature of the many faults that are associated with the dome suggest that the salt was pushing up at the same time that the soft Tertiary sediments were accumulating in the Gulf coastal waters. Many laboratory model experiments using a variety of materials have closely simulated the natural salt domes and their associated structures.

In Louisiana salt is mined from some of these domes. It is very pure, and the reserves are enormous. Texas has many salt domes, but most of the salt is produced in the eastern part of the state from shallow localized deposits that are not domal.

Borax and Borates

No one knows when borax was first discovered, but it is said that Marco Polo introduced it to Europe during the Middle Ages. Sources are known in central and eastern Asia. It was first applied as a soldering flux for the making of silver and gold jewelry. Borax and boric acid, the two most important industrial borates, have since become some of the most versatile chemical minerals known.

Important Boron Minerals

There are over 60 different minerals that contain boron, but only a handful are plentiful or rich enough to be used as commercial sources. With the exception of *sassolite,* that is, natural boric acid, all of these minerals are sodium and calcium borates with considerable water of crystallization. The following table lists these minerals:

IMPORTANT BORON MINERALS

Mineral	Formula	% Boric Oxide
borax	$Na_2O \cdot 2B_2O_3 \cdot 10H_2O$	36.5
kernite	$Na_2O \cdot 2B_2O_3 \cdot 4H_2O$	51.0
colemanite	$2CaO \cdot 3B_2O_3 \cdot 5H_2O$	50.8
ulexite	$Na_2O \cdot 2CaO \cdot 5B_2O_3 \cdot 16H_2O$	43.0
priceite	$4CaO \cdot 5B_2O_3 \cdot 7H_2O$	48.9
sassolite	H_3BO_3	56.3

Adapted from G. A. Connell, *Industrial Minerals and Rocks,* A.I.M.E., Chapt. 7, 1949.

Uses

Borax easily dissolves in water giving a mildly alkaline solution that has antiseptic qualities. It is an excellent cleaning compound, either by itself or mixed with soap, and it also acts as a water softener. The ceramic industry uses much borax and boric acid as fluxes. Porcelain enamels, glazes, pottery, bricks, and many kinds of glasses require boron fluxes. All heat-resistant glass used in the kitchen or laboratory contains boron, which imparts the qualities of strength, brilliance, and durability.

Boron chemicals prevent mildew and mold in starches and serve as preservatives for wood, citrus fruit, leathers, and textiles. Casein needed in the manufacture of paper, plywood, plaster, paints, and calcimines is dissolved in borax solutions. Syrups, pickles, flavor extracts, tobacco, insect repellents, ink, dyes, drugs, and candles are just a few of the great variety of foods and manufactured products that make use of these minerals. Borax is used by the metallurgist in the making of some steels and in the smelting of gold, silver, and copper. Boron is a major constituent of a new and powerful fuel for aircraft.

Boron is a minor but necessary element in the soil for the proper growth of all plants, and so borax is added to fertilizer in small amounts. In greater quantities it is an effective plant killer so that it is used to eradicate noxious weeds.

Production of Boron Compounds

The United States produces nearly all of the boron used in the world. In 1956 over 855,000 tons of boron minerals were mined in the United States. This was 95% of the 900,000 tons world production. Turkey produces boracite (a nonhydrous chloride and borate of magnesium), and Italy mines deposits of sassolite. There are other deposits of boron in the world, but few can compete with the very cheap borates produced from the rich California deposits.

Occurrence of Boron

All boron salts are believed to have been originally brought to the surface of the earth in volcanic vapors. In Tuscany, Italy, steam from volcanic vents and associated hot springs deposit sassolite today. In California there are three kinds of deposits, all associated with playa lakes: (1) bedded deposits in old playas, (2) brines from salt lakes, and (3) surface incrustations around playa lakes. The boron in these deposits probably originated as a result of the extensive Tertiary volcanism that occurred in this region. Thanks to the arid climate and the drainage into playas, the soluble boron salts were retained and concentrated.

KRAMER DEPOSIT

The bedded deposits near Kramer in Kern County, California, were discovered in 1926 in a valley 1 mile wide and 4 miles long. The beds of borax and kernite lie within Tertiary clays and volcanic tuffs at depths between 325 and 1000 feet, and nearly 75% of this interval consists of borates in layers up to 114 feet thick. The beds have been folded, and it is in the most disturbed areas that the kernite is found. Because the Kramer deposits are so extensive, so pure, and so cheaply mined, many other California deposits were abandoned when it was found that they could not compete.

Unlike most borate deposits, these contain no other soluble salts. Because of the depths mining has all been underground, but recently open-cut pits have been developed on shallower occurrences. The borax and kernite are mixed and calcined at the mine, yielding a product with about 45% B_2O_3 that is shipped to the refinery at Wilmington, California.

SEARLES LAKE

The brines from Searles Lake, in San Bernardino County, California (for further description see p. 269) contain about 36% total salts, and 2.84% of this is borax. Two companies working these brines use fractional crystallization to separate sodium and potassium salts as well as bromine and lithium compounds. The borax is a by-product. At another playa, called Owens Lake, brines are similar to those of Searles Lake, but less concentrated. More sodium salts are produced.

DEATH VALLEY

Bedded deposits of colemanite and ulexite were once extensively worked in Death Valley between 1895 and about 1926. Incrustations 1 foot thick of impure borates were common in Death Valley as well as around Searles Lake, Borax Lake, and some playas in Argentina, Chile, and China.

Potash

It has long been known that if one soaked wood ashes in water and then evaporated the water a strong alkali remained. It was formerly common practice to treat the ashes in an iron pot, so that this alkali became known as "pot ash." The alkali formed in this way is potassium carbonate, but the present term, *potash,* is now used to designate the potassium oxide content of minerals. As early as the 1600's settlers in America were extracting potassium carbonate from wood ashes and exporting the product to Europe. This was a natural outgrowth of the process of clearing the dense forests for agriculture, and it became the first chemical industry of the New World.

Now, thick salt deposits and brines, rich in potash, supply the present enormous demands of agriculture and industry.

Potassium Minerals

Of the scores of potassium-bearing minerals the five listed in the following table are most commonly used as sources of potash. They all occur in salt deposits and are more or less soluble in water.

SOURCE MINERALS FOR POTASH

Mineral	Formula	% K_2O
sylvite	KCl	63.2
langbeinite	$2MgSO_4 \cdot K_2SO_4$	27.7
polyhalite	$K_2SO_4 \cdot MgSO_4 \cdot 2CaSO_4 \cdot 2H_2O$	15.6
carnallite	$KCl \cdot MgCl_2 \cdot 6H_2O$	16.9
kainite	$MgSO_4 \cdot KCl \cdot 3H_2O$	18.9

Uses for Potash

Before 1890 most of the potash was used for such purposes as soap manufacture, dyeing, and tanning; in the glass and pottery industry as a ceramic raw material; and as an ingredient in matches, fireworks, and gun powder. Now over 90% is used for fertilizer. It is a vital mineral constituent of the soil that must be replenished regularly for the best yield of field crops and orchards. Potash is normally mixed with other mineral fertilizers in blends containing fixed amounts of potash, phosphorus, and nitrogen. Sugar and starch formation in plants is aided by adequate potash in the soil. It also speeds plant growth and improves the quality of the crop.

Modern industrial demands for potash minerals are illustrated by their use in such diverse processes as the manufacture of baking powder, glues, poison, ammonia, paint, varnish remover, and synthetic rubber. Potassium chemicals are used in photography, as drugs and antiseptics, and as bleaching compounds.

Production of Potash

Before 1861 nearly all potash was made from wood ashes, but in that year potassium minerals were mined for the first time in Germany. By the 1870's the demands of nearly the whole world were supplied by these deposits. Government and private mishandling of this monopoly encouraged extensive exploration for potash minerals in all countries of the world. The embargo of German potash during World War I made a great shortage of the valuable mineral in America, but new low-grade sources were hastily put into production in this country. Germany lost her domination of the industry after 1918 when Alsace was transferred to France. By 1931 new potash production from America, Spain, Russia, and Palestine had started. In 1956 Germany led the world, however, with East Germany producing minerals with a K_2O equivalent of 1,598,000 metric tons. In West Germany the yield was 1,823,221 metric tons. United States deposits were worked for 2,171,584 tons, and France marketed 1,455,000 tons of potash. Spain produced only 256,525 tons.

Geologic Occurrence of Potash Minerals

There are three different modes of occurrence of potash minerals: (1) evaporite marine salt deposits, (2) playa deposits, and (3) brines from salt lakes. Of course, the ultimate source of water soluble potassium salts are the potash silicates such as orthoclase that are so common in nearly all igneous and metamorphic rocks. Weathering of these minerals releases the potash, which is carried to the sea as the soluble potassium carbonate.

GERMAN POTASH

The great German deposits offer an excellent example of the bedded marine salt occurrence. The Permian salt beds occupy a basin that surrounds the uplifts of the Harz, Flechtigan, and Thüringer Wald mountains. Salts are mined in six districts, the best known of which is Stassfurt. The total sequence of evaporites, including halite, anhydrite, gypsum, and potash salts, is nearly 800 feet thick; but considerable local variance occurs as a result of intense folding and faulting. There are three potash zones called the polyhalite, kieserite, and carnallite zones after the main potassium mineral that each contains. In most parts of the 24,000 square miles that are underlain by salt, the potash beds are too deeply buried to mine, but where there are salt domes and anticlinal folds they may be brought to within 1000 to 2500 feet of the surface. It is on these structures that the mines are located.

NEW MEXICO DEPOSITS

The potash salts of the Permian basin mined near Carlsbad, New Mexico, are similar in occurrence to the German deposits except that they have not been so much disturbed by folding. Drilling has proved that at least 3000 square miles are underlain by potash salts, but in the area near Carlsbad the beds are within 1000 feet of the surface along the western edge of the basin. There are many horizons that contain potash minerals. Sylvite is the principal mineral, but polyhalite, carnallite, langbeinite, and others occur in small amounts. Most of the mining is done in the bottom sylvite zone that carries up to 27% K_2O. It is a bed between 5 and 12 feet thick that is largely a mixture of halite and sylvite known as "sylvinite." This is the richest known potash deposit in the world, and it contains reserves estimated at nearly 100 million tons.

SEARLES LAKE

Searles Lake, at an elevation of over 1600 feet in the Mojave Desert of southern California, is a playa that at times has some water at the surface but is usually covered with a crust of soluble minerals. The brine pumped from shallow wells contains a varied assortment of minerals. The lake covers an area of 12 square miles and contains great reserves of soluble salts.

COMPOSITION OF TYPICAL BRINE FROM SEARLES LAKE

KCl	4.70%
NaCl	16.35
Na_2CO_3	4.70
Na_2SO_4	6.96
$Na_2B_4O_7$	1.50
Na_3PO_4	0.16
NaF	0.01
minor constituents	0.30
total salts	34.68
water	65.32
	100.00

From W. A. Gale, *Symposium on Potash*, Ind. Eng. Chem., Vol. 30, 1938.

The lake waters were first processed for borax as early as 1870, but the borax plants were abandoned in 1895 when the Death Valley colemanite deposits first came into production. The difficulties with the German potash producers that started about 1910 and culminated in the embargo of 1915 brought renewed interest in Searles Lake, which was known to contain some potash. It soon became the leading American source of potash, a position held until the New Mexico deposits were in production. Now as many as six or seven valuable salts are refined from the Searles Lake brines.

Nitrates

Nitrogen is a colorless, odorless, and tasteless gas that makes up approximately 78% of the atmosphere by volume. It was first discovered by Scheele in 1772 and was named many years later by Chaptal who recognized its presence in niter. Though nitrogen gas is everywhere, chemical compounds containing nitrogen are not so plentiful and are valued greatly for their many uses. This scarcity is due to the almost chemical inertness of the gas.

There are three significant commercial sources of nitrate chemicals: (1) natural deposits of nitrate minerals, (2) by-product gases, mainly as ammonia (NH_3), from the combustion of coal, and (3) the atmosphere from which the nitrogen can only become useful if it is "fixed," that is, made to combine with some other element. Nearly three-fourths of the world's supply is provided through the "fixation" of atmospheric nitrogen by several different methods. Nitrate minerals and coal supply the remaining amounts.

Uses for Nitrates

All plant and animal life requires nitrogen for the manufacture of proteins. This fundamental mineral substance must always be available in the soil. Legumes, such as clover and alfalfa, contain certain bacteria that have the power to fix nitrogen from the atmosphere. These crops add to the soil's reserves of nitrogen and are used for crop rotation. The alternative to crop rotation is the application of nitrate fertilizers, a use that accounts for nearly 85% of the production of nitrates. Many high explosives such as nitroglycerin, TNT, nitrocellulose, and dynamite are manufactured from nitrogen chemicals. Nitric acid is a chemical basic to many industrial processes and for the manufacture of dyes and some pharmaceuticals. Ammonia is not only an important fertilizer but is a gas commonly used as a refrigerant. Ammonia compounds find varied uses in industry.

Geologic Occurrence of Nitrate Minerals

The sodium nitrate or "Chile niter" deposits of northern Chile are the only natural occurrences of nitrate minerals of economic importance in the world (Fig. 89). They occur in the rainless deserts in a 450 mile belt which is 10 to 50 miles in width. The nitrate minerals act as cementing material in a sandy gravel layer usually a few feet thick and buried by an overburden that varies between an inch or two and up to 40 feet in thickness. The richer highly cemented rock is called "caliche," and it averages about 25% sodium nitrate. About 2 to 3% of potassium nitrate is present along with minor amounts of salts containing iodine, boron, and bromine, some of which can be separated economically. In 1954 over 1,581,000 metric tons of nitrate were produced in Chile.

The climate of the area is exceedingly dry and hot with no rain at all over periods as long as 10 years. The western flanks of the Andes slope gently westward until the land rises again in the coastal ranges. Because this part of South America is in the belt of easterly winds, all of the rain falls on the east side of the Andes. The downward-moving air on the west side is hot and dry over the deserts that lie in the "rain shadow" of the mountains. The nitrate beds are on the far western side of the desert nearest the foothills of the coast ranges. It is generally accepted that the nitrates are carried by ground water that seeps down the Andes towards the Pacific Ocean. In low areas the ground water is closest to the surface and is drawn upwards by capillarity to be rapidly evaporated even before it reaches the surface. This leaves the buried mineral crust that is the caliche nitrate deposits. No one knows how the ground water came to have sodium nitrate and other rare salts dissolved in it. Several theories have been proposed to account for them. Three different organic sources have been suggested: seaweed, bird droppings, and soil bacteria. Lightning in the Andes and the electrostatic charges that accompany the frequent fogs in the area are thought by some to have fixed atmospheric nitrogen which then became dissolved in the ground water. All are possible explanations, but soil bacteria seem the most probable answer in light of the vast area covered by the deposits.

Phosphates

Before the appearance of man there was a natural cycle of phosphorus on the land. Where phosphorus was abundant in the soil, plant life could flourish. Animals ate the plants or ate plant-eating animals, used some of the phosphorus in their teeth and bones, and returned the rest to the soil to start the cycle anew as

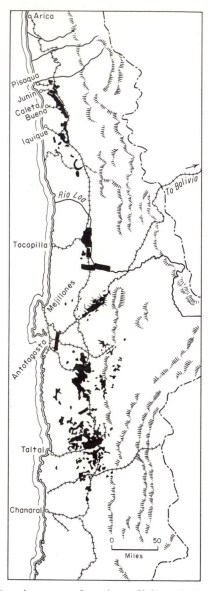

Fig. 89. Map showing the areas of northern Chile underlain by nitrate deposits. (From A. W. Allen, *Eng. and Min. Journ.*, Vol. 126, 1928.   Copyright McGraw-Hill Publishing Co.)

the plants and animals died. Now where crops are removed from the land, the supply of phosphorus in the soil is gradually depleted, and the soil can retain its fertility only where phosphate fertilizers are applied.

Uses for Phosphates

Most phosphate rock is mined for the benefit of agriculture. About 90% of the production is acidulated with sulfuric acid to make the various forms of soluble phosphates that are immediately available for plant growth. A small amount of "raw" phosphate rock is used for soil treatment. The remaining 10% is used to make phosphoric acid and the multitude of industrial phosphate chemicals. Baking powders, water softeners, dental cements, glass, fireproofing compounds, beverages, and photographic chemicals are just a few of the many applications. Elemental phosphorus burns easily with much heat and the generation of dense white smoke. Phosphorus is used in the manufacture of matches (as phosphorus sulfide) and in such ordnance as tracer bullets, bombs, grenades, and artillary projectiles.

Production

About half of the world's production of phosphate minerals is from the United States. In 1957 her yield was 15,600,000 long tons compared to a world total estimated at slightly over 34,000,-000 tons. Extensive deposits in North Africa, particularly in French Morocco and Tunisia, contributed 9,000,000 tons. In the same year U.S.S.R. produced over 5,000,000 tons. The phosphate islands in the Pacific Ocean (Ocean, Nauru, Christmas, and Makatea) shipped 2,000,000 tons.

Geologic Occurrence

Probably the ultimate source of all phosphate material is from the mineral *apatite* $Ca_5(F,Cl)(PO_4)_3$. Apatite is very common in most igneous rocks which make up a large proportion of the crust of the earth. When these rocks weather, the phosphate is made available for plant life in the soil or is carried in solution to the ocean. Here it is used by marine animals in the formation of bones and some kinds of shells. Some marine bacteria even contain phosphorus in their tissues. This organic phosphate and perhaps some that is chemically precipitated eventually becomes incorporated in sedimentary deposits on the ocean bottom. Calcium phosphate occurs in many sedimentary rocks in small amounts, but there were times in the geologic past when phosphates were deposited in higher concentrations in some places. The Mississippian and Permian were such periods.

There are many secondary concentrations of phosphate that are often quite rich and contain large tonnage. Some valuable recent deposits rich in phosphate are called *guano* which is an accumulation of bird or bat droppings. The nesting areas of many islands that serve as rookeries for great numbers of birds become thickly encrusted with guano. Caves that harbor swarms of bats during the day often contain rich guano deposits.

FLORIDA

Florida is the leading state in the production of phosphate. Most of the deposits lie in an area 60 miles long and 30 miles wide, just east of Tampa. The area is underlain by a Miocene phosphatic limestone. Weathering and solution of the soluble calcite from the limestone have left a residual mantle of sand- to pebble-size particles of insoluble calcium phosphate. This loose, easily mined surface deposit is called the "land pebble phosphate." The particles average ½ inch in diameter and contain between 66 and 70% BPL.[1]

The so-called hard-rock phosphates are farther from the coast in Pliocene rocks. The deposits contain many great boulders of phosphate weighing several tons each. It is believed that these masses were deposited in limestone as replacement bodies and cavity fillings from phosphate that was leached out of beds that once lay above. Now that the limestone is weathered away, the insoluble phosphate masses remain. When cleaned from the surrounding clay, the material contains 70 to 78% BPL.

It is fortunate that the Florida deposits are close to the cotton and tobacco fields of the South. Much fertilizer is required for the growth of these soil-depleting crops.

WESTERN STATES

Rocks rich in phosphate crop out in Montana, Idaho, Utah, and Wyoming (Fig. 90). Some phosphate occurs in a few Mississippian

---

[1] The content of phosphate is always given as tricalcium phosphate, which is referred to as "bone phosphate of lime," hence BPL.

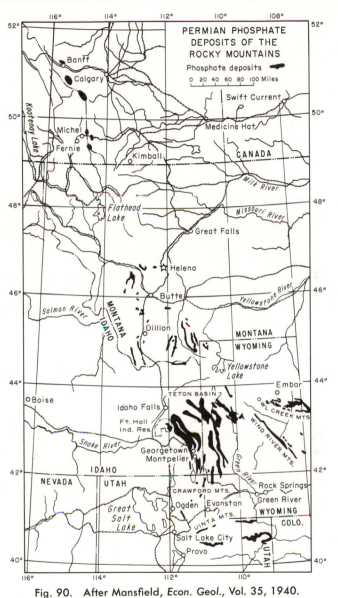

Fig. 90.   After Mansfield, *Econ. Geol.*, Vol. 35, 1940.

beds, but the Phosphoria formation of Permian age contains all of
the commercial deposits.   The producing horizons are shale beds
that are 3 to 4 feet thick, but at Conda, Idaho, two shales are
mined, each about 7 feet thick with BPL content between 60 and
70%.   The rocks have been folded and faulted to such an extent

that underground mining is necessary. It is interesting that the phosphate rock contains between 0.001 and 0.065% uranium oxide and up to 1.2% vanadium oxide. One reason for the development of these phosphates is the availability of sulfuric acid that is a by-product from the copper, lead, and silver smelters in Montana and British Columbia. Available markets for the product are far away, so that the cheap acid source makes it possible for the deposits to compete with those from the Southeast. The reserves of phosphate rock in these states is enormous.

RUSSIA

The great Russian phosphate deposits of the Kola Peninsula occur in a large intrusion of an unusual igneous rock that contains concentrations of apatite as rich as 65 to 70%. The main producing area has a body of ore over 600 feet thick and extending for several miles. Open quarrying is done in the summer, and underground operations produce the phosphate during winter months. Crushing and flotation give a concentrate which is nearly 88% BPL.

Sulfur

Sulfur was well known even in ancient times when it was referred to as "brimstone." The alchemists thought of brimstone as the "principle of combustion" because of its ability to burn.

Sulfur is a familiar nonmetallic element with a characteristic yellow color. It is soft yet brittle and melts at a relatively low 113°C. It burns with a blue flame and liberates noxious sulfur dioxide gas.

Uses for Sulfur

Sulfur is one of the mineral substances most widely used by modern industry. Nearly half of the production goes for the manufacture of sulfuric acid. Over a third of this powerful acid is employed in the formation of superphosphate fertilizers, and much is used to manufacture a host of sulfur-bearing industrial and pharmaceutical chemicals. Sulfuric acid is consumed in large amounts for certain processes in the refining of petroleum. Explosives, paints, many artificial textiles, and several metallurgical processes require sulfuric acid.

Powdered sulfur is used in agricultural fertilizers, insecticides, and fungicides. In paper plants sulfur dioxide fumes are dissolved in a milky lime and water mixture to produce a solution that is able to digest wood chips, leaving only the desired cellulose fiber. All rubber products contain some sulfur, between as much as 30% in hard battery rubber and as little as 1.5% in rubber used for tires. Synthetic rubber could not be made without sulfur chemicals. Dyes, explosives, some food preservatives, fumigants, and many other minor uses add to the great demand for sulfur.

World Production of Sulfur

Prior to 1904 most of the sulfur used in the world came from the great deposits in Sicily. In that year the mining of sulfur commenced in Louisiana using the newly perfected Frasch process (see p. 280), and by 1916 the United States assumed the lead in the production of this vital mineral. In 1956 the American deposits of native sulfur yielded 6,484,285 tons, over 81% of the 8,000,000 tons produced in the whole world. In recent years Mexico has become the second leading producer with 758,415 tons in 1956. Italy marketed 170,000 tons and Japan 243,312. Chile, Argentina, and France had minor production.

Sulfur extracted from pyrite (FeS_2) amounted to almost as much tonnage as sulfur produced as the native element. This mineral, the so-called fools gold, is mined in great quantities, usually as a by-product in ores containing copper, gold, or other metals, but sometimes as an ore of sulfur alone. Sulfur from this source amounted to 6,700,000 tons in 1955. Japan and Spain led in the production of pyrite sulfur with 1,136,000 and 1,100,000 tons respectively. Italy, United States, Cyprus, and Norway were also major producers.

Geological Occurrences of Native Sulfur

There are three different kinds of native sulfur deposits of economic importance: (1) sulfur in the cap rock of salt domes, (2) sulfur in sedimentary beds, and (3) deposits of sulfur in regions of volcanic activity.

The origin of sulfur in the cap rock of salt domes is as much a problem as the origin of salt domes themselves. Most investigators agree that the sulfur is derived from the mineral anhydrite, a common impurity in the salt. A layer of anhydrite usually lies directly upon the salt and is the bottom layer of the cap rock.

The sulfur commonly impregnates a porous limestone that lies directly upon the anhydrite. Petroleum or bituminous residues are nearly always present. According to a well-accepted theory, anaerobic bacteria,[2] known to occur in great numbers in oils and oil-field water, used for their life processes the carbon from petroleum and the oxygen from the calcium sulfate. This changed the anhydrite to calcium sulfide, a substance that reacts with water and carbon dioxide (also produced by the bacteria) to give calcite and hydrogen sulfide. Hydrogen sulfide is a gas that easily changes to sulfur and water whenever it encounters oxygen. Thus calcite limestone and sulfur are the only solid products of these reactions that are given below in a simplified fashion:

$$(1) \qquad \underset{\text{anhydrite}}{CaSO_4} + \underset{\text{petroleum carbon}}{2C} + bacteria \longrightarrow$$

$$\underset{\text{calcium sulfide}}{CaS} + \underset{\text{carbon dioxide}}{2CO_2} + more\ bacteria$$

$$(2) \qquad CaS + CO_2 + H_2O \longrightarrow \underset{\text{calcite}}{CaCO_3} + \underset{\text{hydrogen sulfide}}{H_2S}$$

$$(3) \qquad H_2S + \underset{\text{oxygen}}{O} \longrightarrow H_2O + S$$

A similar origin is proposed for the sulfur that is found in sedimentary beds. The deposits are thought to have formed by bacterial action in saline and stagnant isolated basins. Such basins today are so saturated with hydrogen sulfide that marine life cannot live.

Gases given off from volcanoes commonly contain sulfur vapor, sulfur dioxide, and hydrogen sulfide. The hydrogen sulfide can oxidize and deposit sulfur in the cavities of porous volcanic rocks, or it may be carried in solution in spring waters where bacteria cause the deposition of sulfur. A reaction between sulfur dioxide and hydrogen sulfide also yields sulfur. Vapors of sulfur merely deposit solid crusts of crystals where they cool at the mouths of gas vents.

GRANDE ECAILLE MINE

Of the more than 300 salt domes that are known in the Texas-Louisiana coastal region only 14 contain commercial deposits of

---

[2] Anaerobic bacteria are one-celled plants that do not require air. They have the ability to extract carbon and oxygen from certain chemical compounds on which they live.

sulfur (Fig. 91). One of these is the Lake Washington salt dome located on the Mississippi delta some 43 miles south of New Orleans. The region is so swampy that all transportation is by boat along canals. Monumental engineering problems were solved to erect the plant and to provide fresh water for mining operations (Fig. 92) under such conditions.

The cap rock of the salt dome lies at a depth of 1200 feet and covers an area of about 1100 acres. A typical cap-rock section has about 100 feet of barren limestone at the top, underlain by some 200 feet of sulfur-bearing porous limestone. Beneath this there is about 100 feet of anhydrite and gypsum which rest upon the salt. Only about 200 acres of the cap rock have been worked for sulfur since 1934 when production was started by the Freeport Sulphur Co. The mining is by the Frasch process, and enough sulfur remains that the operations can continue at the present rate of

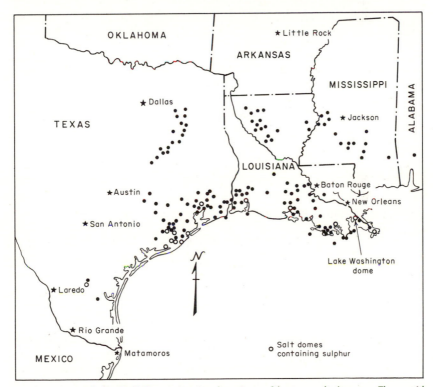

Fig. 91. Map of the Gulf Coast showing locations of known salt domes. Those with sulfur are specially designated. (Modified from Landes, *Petroleum Geology*, John Wiley and Sons, 1951.)

Fig. 92. Aerial view of sulfur wells on the Grande Ecaille deposit. Note swampy nature of area and large blocks of sulfur stored in left background. (Photo courtesy Freeport Sulphur Company.)

production for another 80 years or more. There are many rich oil pools in the cap rock and around the periphery of the salt dome.

New salt dome deposits in Mexico are producing heavily, and a recent discovery offshore near Grand Isle, Louisiana, will someday be important.

Mining of Salt Dome Sulfur

The Frasch process is used almost exclusively for the mining of sulfur from the cap rocks of salt domes. The process was patented by Herman Frasch in 1890, but it was not until 1904 that it was first used for commercial production. This method makes use of the low melting point of sulfur and the generally porous nature of the limestone in which it occurs. Water is heated to 350°F and is forced down a well into the sulfur horizon. The sulfur melts, and being twice as heavy as water sinks to the bottom part of the limestone and the hot water remains on top. Only the molten sulfur is brought to the surface and the excess water is drained off from "bleeder wells" at the edge of the field. Figure 93 shows the ingenious pipe arrangement of the sulfur well.

The molten sulfur is usually piped to huge vats where it is spread out on the surface within forms until a block is built up of cooled solid sulfur several hundred feet square and about 60 feet thick. This is broken by power shovels and shipped when needed. No further treatment is normally required as the sulfur from the wells is between 99.55 and 99.9% pure. It is now possible to ship molten sulfur for great distances in covered barges. The thin crust of solid sulfur which forms on the surface is excellent insulation for the remaining molten sulfur.

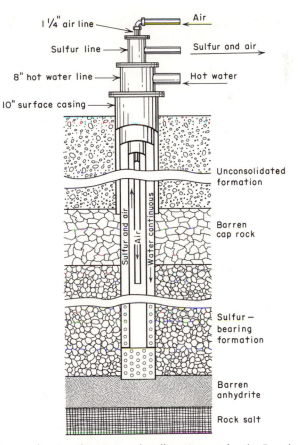

Fig. 93. Diagram showing the pipe and well equipment for the Frasch process for the "mining" of sulfur from the porous cap rocks of salt domes. (From Lundy, *Industrial Minerals and Rocks*. Copyright A.I.M.E., 1949.)

SULFUR DEPOSITS OF SICILY

These rich and historical sedimentary deposits of south central Sicily lie in structural basins that are 1 to 5 miles long and up to 3000 feet in width. The sulfur-bearing limestones reach a maximum thickness of 200 feet and contain between 12 and 50% sulfur. These beds lie below a sequence of sandstones, marls, clays, and gypsum. Bituminous shales are interbedded with the sulfur layers. The rocks have been folded and faulted.

Mining carried out by underground operations is modern and efficient, in the larger mines at least. There is a constant hazard of fire because of the accumulations of combustible hydrogen sulfide gas. The sulfur was once extracted by the *Calcaroni method* in which the ore was stacked in piles 15 to 100 feet in diameter and up to 15 feet deep. The piles were then ignited, and about half the sulfur burned to melt out the remaining half that was recovered. Now masonry kilns are employed, and the recovery is between 80 and 85%, the rest of the sulfur acting as fuel.

JAPANESE SULFUR

The main deposits lie on northern Honshu and Hokkaido islands. There are more than 40 separate occurrences of sulfur, most of which are fissure fillings and impregnations of porous volcanic rocks on the numerous volcanoes. Sulfur-bearing clays and volcanic ash are also productive. Most deposits are small, and 12 mines are responsible for most of the production.

SELECTED REFERENCES

GENERAL

Johnstone, S. J. *Minerals for the Chemical and Allied Industries,* 692 pp., John Wiley and Sons, New York, 1954.

SALT

Looker, C. D. Salt as a chemical raw material, *Chem. Ind.,* pp. 594–602, Nov. 1941; pp. 790–799, Dec. 1941.

Nettleton, L. L. History of concepts of Gulf Coast salt-dome formation, *Am. Assoc. Pet. Geol. Bull.,* Vol. 39, No. 12, pp. 2373–2383, 1955.

Phalen, W. C. Salt resources of the United States, *U. S. Geol. Surv. Bull.* 669, 1919.

U. S. Geological Survey. *Salt deposits of the United States, incomplete compilation from published sources, 1950,* U. S. Geol. Surv. Open File Rept. No. 235, map, 1954.

BORAX

Noble, L. F. Borate deposits in the Kramer District, Kern County, California, *U. S. Geol. Surv. Bull.,* 785, 1926.

Santmyers, R. M.   *Boron and its Compounds,* U. S. Bur. Mines Inf. Circ. 6499, 37 pp., 1931.

Vonsen, M.   The discovery of borates in California, *The Mineralogist,,* Vol. 3, pp. 3–4, 21–25, 1935.

NITRATES

Johnson, B. L.   *Nitrogen and its compounds,* U. S. Bur. Mines Inf. Circ. 6835, 1931.

Wright, C. W.   Chile mineral resources, *Foreign Minerals Quarterly,* Vol. 3, No. 2, 1940.

POTASH

Johnson, B. L.   *Potash,* U. S. Bur. Mines Econ. Paper 16, 1933.

Smith, J. P.   Notes on the geology of the potash deposits of Germany, France and Spain, *Min. Eng.,* Vol. 187, No. 1, pp. 117–121, 1950.

Turrentine, J. W., et al.   *Symposium on potash,* Ind. Eng. Chem., Vol. 30, No. 8, 853 pp., 1938.

Turrentine, J. W.   *Potash in North America,* Amer. Chem. Soc. Mon. Series, Vol. 91, Reinhold Publishing Co., New York, 1943.

PHOSPHATE

Cathcart, J. B. (and others).   *The Geology of the Florida land-pebble phosphate deposits,* U. S. Geol. Surv. Open File Rept. No. 177, 21 pp., 1953.

Harris, R. A., D. F. Davidson, and B. P. Arnold.   Bibliography of the geology of the western phosphate field, *U. S. Geol. Surv. Bull.* 1018, 88 pp., 1954.

McKelvey, V. E., and others.   *Domestic phosphate deposits,* U. S. Geol. Surv. Open File Rept. No. 178, 49 pp., 1953.

Mansfield, G. R.   Phosphate deposits of the United States, *Econ. Geol.,* Vol. 35, pp. 405–429, 1940.

Mansfield, G. R.   Phosphate resources of Florida, *U. S. Geol. Surv. Bull.* 934, 1943.

SULFUR

Banfield, A. F.   Volcanic deposits of elemental sulphur, *Can. Min. and Met. Bull.* Vol. 47, No. 511, pp. 769–775, 1954.

Espenshade, G. H., and C. H. Broedel.   *Annotated bibliography and index map of sulphur and pyrite deposits in the United States and Alaska* (with references to July 1, 1951), U. S. Geol. Surv. Circ. 157, 48 pp., 1952.

Haynes, W.   *The Stone That Burns,* D. Van Nostrand Co., Princeton, 1942.

Lundy, W. T.   *The development of the Grande Ecaille salt dome,* A.I.M.E. Tr. 109, 1934.

Ridgeway, R. H.   *Sulphur,* U. S. Bur. Mines. Inf. Circ. 6329, 1930.

gems and gemstones

Even before man had learned to use metals, he prized decorative stones. Archaeologists tell us that man used jade as early as 22,000 B.C., and turquoise was introduced sometime between 7000 and 3400 B.C. All of the gems of ancient times were found in gravels, but some turquoise was mined from hard rock in the Sinai Peninsula.

Almost from the very first, man learned to polish, shape, and carve colorful gemstones. Seals with ornamental carvings are well represented by the Egyptian scarabs and the Greek gems of the first millenium before Christ. They were executed in great detail and with superb artistry. Eventually gemstones were appreciated for their own beauty, and gem carving became less common. The rounded lozenge-shape called a "cabachon" was once the most widely used, and finally in the Middle Ages, the art of faceting was introduced. Now nearly all clear precious and semiprecious gemstones are faceted, but the cloudy, transluscent, and opaque materials are still cut and polished as cabachons.

MINERALS USED FOR GEMSTONES

Any uncut mineral substance that man considers beautiful may be classified as a gemstone. After it has been cut it is called a "gem." Let us consider what properties determine why any one mineral is a gemstone.

Beauty is, of course, the first and primary quality that every gemstone must possess. Values of beauty sometimes change through the ages as dictated by fashion, but for the most part gemstones highly prized 3000 years ago are considered just as lovely today. The beauty of a gemstone may lie in many or all of the following special properties: (1) attractive color, (2) brilliance or luster that is outstanding, (3) beautiful and unusual optical effects, such as the "stars" in sapphires and rubies and the play of colors displayed by precious opal, and (4) the high polish or delicate carving the stone may take.

Color was once the determining factor of a gemstone's worth. In ancient times a brilliant blue sapphire and a beautiful purple amethyst might have been considered of equal value. With faceted gems, the brilliance of the sapphire becomes apparent in contrast to the duller, glassy amethyst. The sapphire has a higher *index of refraction*,[1] that is, the greater ability to bend light. If a gem bends the violet wave lengths of light to a greater extent than it bends the red wave lengths, it is said to have *dispersion* and shows flashes of color from the light passing through it. Diamond has both a high index of refraction and a high dispersion giving the familiar brilliance to this gem. Stars in sapphires and rubies are caused by netlike inclusions of fine crystals that give the gem a cloudy appearance. The rainbow flashes of color in opal are caused by fine cracks or by lenses of opal with a slightly different index of refraction.

Durability is an important quality for a gemstone. Most of the precious and semiprecious stones are hard and tough. They can be worn in rings and necklaces, constantly subjected to bumps, dust, and abrasion without showing any appreciable wear. In the mineralogist's scale of hardness, diamond, the hardest natural substance in the world, is given the value of 10. Talc, a very soft

---

[1] The index of refraction of any transparent substance may be considered as the ratio of the velocity of light in air to its velocity in the substance. The index is measured by the amount of bending of a light ray as it passes from air into the substance.

mineral, has the value of 1. Steel and window glass are about 5½. Nearly all gemstones are 6 or above, but pearls are only hardness 3 and must be carefully treated lest they become scratched. Minerals that are excessively brittle or ones that cleave too easily do not make good gems. The durability of gems is demonstrated by the many perfect specimens that are thousands of years old yet as beautiful as the day they were first cut.

Rarity, of course, makes anything valuable, and this is particularly true of gemstones. Emerald is far more valuable than sapphire because not nearly as many emeralds can be found. Diamonds are also rare, but the marketing of the gem-quality stones is now carefully controlled so that the value is kept relatively constant.

The following table lists the important minerals that are used for precious and semiprecious gems. Some of these are discussed in the subsequent pages of this chapter.

Diamonds

The diamond is the most desired of all the precious gems, and commerce in diamonds is far greater than for any other gem material. Aside from its beauty and value as a gemstone, the diamond ranks as one of nature's most unusual minerals. Chemically it is pure carbon, the same element that makes up graphite and is so important in coal and coke. Diamonds will burn, but only when they are heated to high temperatures in oxygen. At normal temperatures the diamond is so inert that the strongest acids and alkalis do not affect it. Diamond crystals are common and usually appear in octahedral forms (Fig. 94). The mineral cleaves in exceedingly smooth planes that are parallel to the crystal faces. Diamond cutters make use of these cleavage directions in the initial trimming of uncut stones. It is the hardest natural substance and can easily scratch any other mineral or metal. Only another diamond can scratch a diamond, so diamond dust is used in the grinding and polishing of these gems. The great hardness accounts for its durability and for the value of industrial diamonds. Colorless diamonds are highly valued, whereas those with tints of yellow, gray, or brown are not considered as good. Deeply colored diamonds in blue, green, or red are rare and are treasured as the

IMPORTANT GEMSTONES *
Precious Stones

Stone	Chief Constituents	Color	Hardness	Main source
diamond	C	colorless to yellow	10	South Africa, Brazil
emerald (beryl)†	Be, Al, Si, O	green	7.5–8.5	Colombia, Egypt
ruby (corundum)	Al, O	red	9	Burma, Ceylon
sapphire (corundum)	Al, O	blue	9	Ceylon, Burma, Thailand
precious opal	Si, O	variegated	5.5–6.5	Australia, Hungary, Mexico

Semiprecious Stones

Stone	Chief Constituents	Color	Hardness	Main source
amethyst (quartz)	Si, O	purple	7	India, Iran, Brazil
beryl	Be, Al, Si, O	various	7.5–8.5	U.S., Africa, Brazil
benitoite	Ba, Ti, Si	deep blue	6.5	California
chrysoberyl	Be, Al, O	green, yellow	8.5	Europe, Madagascar
feldspar	K, Na, Ca, Al, Si	various	6	worldwide
garnet	Al, Fe, Mg, Si	red, green	6.5–7.5	Arizona, S. Africa, Ural Mtns.
jade—nephrite	Ca, Mg, Fe, Si	green to white	5.5	Burma
jade—jadeite	Na, Al, Si	green	6.5	Burma, China, Wyoming
kunzite (spodumene)	Li, Al, Si	lilac	6.5–7	California, Madagascar
lapis lazuli	Na, Al, S, Si	dark blue	5–5.5	India, Greece, California, Siberia
peridot (olivine)	Mg, Fe, Si	olive green	6.5–7	Levant, Egypt, Burma
quartz	Si, O	various	7	worldwide
spinel	Mg, Si	reddish	8	Ceylon, Burma, Thailand
topaz	Al, F, Si	yellowish	8	Brazil, Ceylon, Montana
tourmaline	Al, Fe, B, Si	green, pink	7–7.5	Urals, Madagascar, California, Maine
turquoise	Al, P, O, H	blue	5–6	New Mexico, Iran, Turkestan, Egypt
zircon	Zr, Si	red, orange	7.5	Ceylon, Thailand

* After A. M. Bateman, *Economic Mineral Deposits*, 2nd ed., Chapt. 24, John Wiley, 1950.
† Common mineral name given in parentheses.

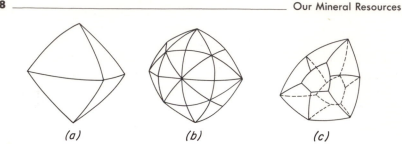

(a)   *(b)*   *(c)*

Fig. 94. Common crystal forms of diamonds.

most valuable of all gemstones. The specific gravity of 3.52 is notably higher than that of the common rock minerals and accounts for concentrations of diamonds in placer deposits. The very high index of refraction and its dispersion power, greater than in any other gem, give the familiar "fire" and brilliance to the diamond.

Uses for Diamonds

Not only is diamond the most prized of all gems, but it is an important and useful industrial mineral because of its superior hardness. About three-quarters of the yearly production of diamonds is classified for industrial uses. Industrial diamonds are divided into four categories: (1) *fine industrials,* (2) *bort,* (3) *ballas,* and (4) *carbonado.*

The *fine industrial* stones are single crystals of nearly gem quality, but because of poor color they cannot be sold as gem stones. These are generally shaped as special cutting tools or made into wire drawing dies. *Bort* consists of diamond crystals that are misshapen and have many flaws and inclusions. Bort is wholly unsuited for gem purposes but is widely used for rock-drilling bits (Fig. 95). These diamond drill bits cut through even the hardest rocks and are most used where rock cores are needed to sample some buried ore. Some bort is crushed to give different grades of diamond abrasive powder used in special grind wheels or for the cutting of gem diamonds. Rock saws have diamond grit pressed into the edges or bonded into special castings. Both the *ballas* and the *carbonado* are fine-grained diamond aggregates. The ballas is a spherical aggregate with crystals arranged radially, and the carbonado is a black dense mass of fine-grained diamonds. Both are very tough and hard, and neither have any cleavage. They are excellent for special lathe tools and rock bits.

Geologic Occurrence of Diamonds

The only primary sources for diamonds are the cylindrical carrot-shaped intrusions of ultrabasic igneous rocks called pipes (Fig. 96). The pipes consist of a rock known as kimberlite in which the diamonds occur as well-developed crystals, crystal aggregates, and fragments. It is generally held that the diamonds crystallized from the magma at great depth and were carried upwards when the magma invaded the surrounding rock. In the DeBeers mine two halves of a single rod-shaped crystal were found at different levels showing that there must have been some violent vertical movement after the diamonds had crystallized.

The great hardness, the high specific gravity, and the chemical inertness of the diamond make it an ideal mineral for concentration in placer and residual deposits. Long before anyone had ever found a primary source for diamonds, the gravels of India and Borneo were supplying the world's demand for this gem. Now stream deposits in South Africa and Brazil yield about 60% of the total diamond production.

Both the primary and secondary sources for diamonds are illustrated by the following examples:

THE PREMIER PIPE

The Premier pipe is one of the largest pipes, with a diameter of 2800 feet. The typical greenish-black kimberlite is in the form of a vertical funnel-shaped plug which has penetrated a sequence of shales, lava flows, and metamorphic rocks ranging in age between

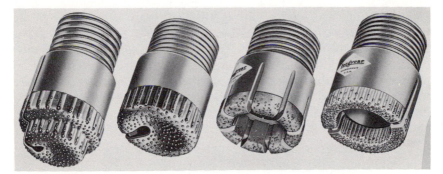

Fig. 95. A variety of diamond rock-drilling bits. The two on the right are designed to take rock core samples. (Photo courtesy E. J. Longyear Company.)

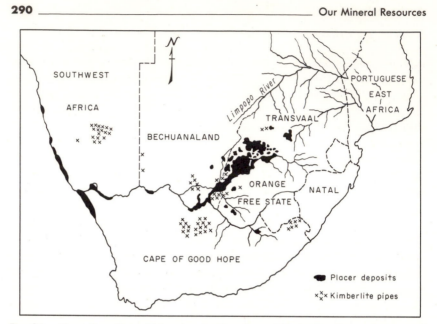

Fig. 96. Map of southern Africa showing locations of kimberlite diamond pipes and areas of placer diamond accumulations. (From Lilley, *Economic Geology of Mineral Deposits,* Henry Holt and Co., 1936. Based on maps from A. F. Williams and O. Stutzer.)

Mesozoic and Precambrian. The outcrop was marked by a yellow decomposed kimberlite soil that extended to depths of over 70 feet. This was the easily mined discovery ore called "yellow ground" that changed at depth to the hard unaltered rock once called the "blue ground." Much of the *kimberlite* has been fractured and re-cemented by successive intrusions of the molten magma and is called a "breccia." In the kimberlite are many inclusions of the rocks through which the pipe has penetrated. The ore averages 16 to 19 carats per 100 loads (1600 pounds each) of ore, which comes to about 0.0000052% of diamond by weight. Over the years of production the average ore content has been about 0.2 carat per ton. It is small wonder that diamonds are rarely seen in the diamond ore.

LICHTENBURG ALLUVIAL FIELD, TRANSVAAL

In 1926, these richest of all placer diamond deposits were dis-covered in an area about 100 miles west of Johannesburg. Soon after they became the scene of a great rush for claims in which

25,000 "runners" were said to have competed. For the next 4 years the production from these deposits was outstanding. In 1927, for example, 2,100,000 carats were produced, nearly as much as from all the pipes in the same year. Williams (1932) and Bateman (general reference) describe the placer deposits in detail.

BRAZILIAN PLACER DIAMONDS

The first diamonds were found in gold-bearing gravels of Minas Gerais, Brazil, about 1720. Subsequently, other placers were discovered in the states of Bahia, Matto Grosso, and Goyaz. Some fine gemstones have been recovered, and the Bahia fields are the world's major source of the excellent carbonado industrial stones. Over 150,000,000 dollars worth of diamonds have been marketed.

Most of the production has been from gravels in present stream channels and old gravel terraces. The source of the diamonds in these modern placers is thought to be the Lavras beds, which are partly metamorphosed gravels that lie on an eroded surface of Precambrian metamorphic rocks. These gravels are probably early Cambrian or Precambrian age and seem to have had a glacial origin. Some diamonds have been found in the old gravels, and the best modern placer deposits of Minas Gerais and Bahia occur where the streams cut the Lavras beds. A breccia with a claylike matrix cementing angular blocks of sedimentary rock contains the diamonds in the Anga Sija deposit of Minas Gerais. It is the kind of material that would result from the weathering of a rock like kimberlite, but no diamond-bearing pipes have been found in Brazil.

Ruby and Sapphires

Both ruby and sapphire are clear colored varieties of the mineral corundum. This mineral commonly forms hexagonal crystals that are barrel shaped. Because it has a hardness of 9 and no cleavage, it is very durable. Diamond is the only natural mineral that is harder. Corundum is chemically aluminum oxide, a substance exceedingly stable and resistant to weathering. Common corundum is usually opaque or transluscent and is gray, brown, or black in color. Transparent crystals of all colors are considered of gem quality. Those that are deep red are rubies, and the blue corundum is the true sapphire. Green, yellow, purple, bluish-gray, pink, and colorless varieties are also called sapphires, but the color term always prefixes the name. A common but misleading practice is to use the word *oriental* for sapphires that have the color of other gems, that is, *oriental emerald* (green), *oriental amethyst* (purple), and *oriental topaz* (yellow). The best rubies have a deep purplish-red, called pigeon's-blood red. A cornflower blue, called *Kashmir blue,* is considered the best color for sapphires.

For gems with the deepest and best colors, rubies and sapphires should be cut with the top of the gem parallel to what was the end of the crystal. When viewed through its side the ruby crystal is a much paler red and the sapphire appears to have a yellowish-blue tint. This is caused by an optical property called *dichroism* in which a mineral absorbs different wave lengths of light passing in different directions through the crystal.

Both rubies and sapphires commonly have a kind of fibrous structure due to inclusions arranged in a latticelike pattern. This gives them a cloudy appearance, and when cut as cabochons they may exhibit a six-pointed white star of light that seems to be suspended within the gem. These star sapphires and star rubies have become very popular in recent years.

Most of the best rubies are mined in northern Burma near the town of Mogok. Here rubies, sapphires, spinels, and tourmalines occur in an ancient limestone which has been intruded by two different igneous masses. The contact metamorphic marble itself is too lean to work, but in solution cavities, caves, and stream gravels in the area there have been local secondary concentrations of these gems. An old lake bed in the region has a zone of gem-bearing pebbly clay now buried by some 15 feet of overburden. Of course, the solution and weathering of the contact marble gave rise to all of these deposits of secondary concentration. The area has been an important source of ruby and sapphire since the fifteenth century.

The most beautiful deep-blue sapphires are mined from a pegmatite dike in the Zanskar Range in Kashmir. This primary deposit was first discovered in 1908, and since that time several million carats have been produced. Another important source of sapphires is in southern India near Mysore and Madras, where the stones are recovered from gravel deposits. Similar deposits occur in Siam, Cambodia, and Anakie, Queensland. The Queensland gravels lie above the present valley and yield deep inky-blue sapphires that appear almost black in artificial light. Ceylon has been a source for sapphires and some rubies for the last 2500 years.

In Yogo Gulch, Fergus County, Montana, pale steely sapphires and some rubies have been mined from a dike of unusual composition. About a quarter of the stones from this area were of gem quality, and hundreds of thousands of carats were produced. There has been little activity here since 1929.

Much of the value of rubies and sapphires has been in their use for jewel bearings in watches and delicate instruments. Most of the small and off-colored stones were used for this purpose. This market vanished when the process for making synthetic rubies and sapphires was perfected (see p. 297).

Emeralds and Other Beryls

The mineral *beryl* is an aluminum silicate of the light metal, beryllium (see p. 193) and its only important ore mineral. Crystals of beryl are always simple hexagonal forms, and some are known that weigh several tons. It is a mineral of great durability because of its superior hardness of 8 and its absence of any good cleavage. However, because its specific gravity is only 2.7, it is rarely found in any secondary placer concentrations. Gems of beryl have little brilliance or fire, because the mineral has a low index of refraction and low color dispersion power. They must all depend upon striking colors for their beauty and value. There are at least six different gem varieties of beryl.

Emerald (green beryl) is probably the most valuable of all gems. Large flawless ones are unknown, and slightly flawed stones of moderate size are exceedingly rare and may cost up to $10,000 per carat. It is common for the color to be variable and for the stones to contain feathery inclusions. The best emeralds are a deep grass green and have a kind of velvety appearance.

Aquamarine is a bluish-green variety of beryl with a lighter shade than emerald. Large unflawed aquamarines are relatively common so that the stone commands a much lower price than emerald. One huge crystal of aquamarine was found in Brazil not so long ago that weighed 243 pounds and was completely transparent. It sold uncut for $25,000. Greenish aquamarines can be heat-treated to 450°C, a process that changes the color to the more highly prized bluish tint.

Golden beryl and *heliodor* are yellow varieties of beryl, the latter being a shade that is found only in South-West Africa.

Morganite is the gem variety of beryl with a pink or rose color.

It was named for J. P. Morgan, who was noted among other things as an avid collector of gems. The color of morganite gems may also be improved by heat treatment.

Goshenite is the name for the clear and colorless variety of beryl.

Nearly all of the best emeralds come from either the *Muzo* or the *Chivor* mines in Colombia. At Muzo, veins of calcite, which must have formed at very high temperatures, cut highly folded black shales. The emeralds occur in crystal clusters lining cavities and are accompanied by a variety of unusual minerals. The Chivor deposits are much like those at Muzo except that the associated minerals are not the same. The Colombian government owns both mines, and they are operated only sporadically on concession. In the Murchison Range, Transvaal, and in the Ural Mountains near Takovaya, Russia, emeralds have been mined from schists near the contact zones of pegmatite intrusions. It was deposits of the same kind that were worked over 2000 years ago near the Red Sea in Egypt. Pegmatites and altered marble have yielded some emeralds in Brazil.

Morganite is mined in San Diego County, California, and also in Madagascar. It occurs in pegmatites associated with pink tourmaline and the pink spodumene gem called *kunzite*. Aquamarine is recovered from pegmatites in Madagascar, Ireland, the Urals, the island of Elba, and in the United States in Maine, Connecticut, North Carolina, Colorado, and California. Golden beryl is a product of Bahia, Brazil, and is also found in Ceylon, Madagascar, Maine, and Connecticut. *Goshenite* was named for, and is produced mainly from, Goshen, Massachusetts.

Opal

The ancients once considered opal as one of the most valuable of gems, second only to emerald. Unfortunately, during later years opal fell into disfavor because of a superstitious belief that an opal brought nothing but disaster and unhappiness to its owner. Only in recent years has this strange notion been dispelled, and opal is again prized as a lovely gemstone. The name was taken from the Sanskrit word, *upala,* which means precious stone.

The opal is the only common gemstone that is not crystalline but an *amorphous* substance. This means that the atoms that make up the mineral are not arranged in any orderly fashion. Opal, like other *amorphous* minerals, forms from the hardening and drying of gelatinous material. Some of the water is retained, and opal may have as little as 2% or as much as 13%. Opal is mainly silica. It is a gemstone with an unusually low specific gravity (1.9 to 2.3) and a hardness of 5.5 to 6.5. It is not a durable stone because of its inferior hardness and brittle nature. Opal easily absorbs stains and oils, which can discolor it and even ruin its gem properties.

The beauty of precious opal lies in its ability to give off flashes of color (opalescence) in reflected light. It is thought that the play of colors is due to a multitude of fine cracks caused by shrinkage of the gel as the opal formed. An optical effect, called *interference,* results in a rainbow color when light is reflected back and forth across the thin opening. The same color effects occur even if the cracks are filled with opal of slightly different index of refraction. Both common and precious opal may be of any color, but only the precious varieties show the colorful *opalescence.*

Of the many gem varieties of opal, black opal is considered most valuable. It is a deeply colored stone that also has the flashes of color from within. White opal is a milky variety, and fire opal is a more transparent type with a lovely yellow, orange, or red color. *Girasol* is a cloudy blue-white opal. A great assortment of common opal, without opalescence, include some with the descriptive names *milk opal, agate opal, rose opal, moss opal, opal jasper;* and many others. These are sometimes cut as semiprecious stones.

Most opal occurs in regions of recent volcanic activity where the still-hot solutions are able to decompose volcanic glass and vulnerable silicate minerals. The solutions carry the resulting silica in a gelatinous state, and this material coagulates in cracks and bubble holes in the newly formed volcanic rocks.

Australia is the main source of opal today. In New South Wales black opal occurs as fillings in cavities and narrow seams in sandstone down to depths of 100 feet. Queensland and South Australia have similar deposits. An early but still productive source for precious opal is near Eperjes in Czechoslovakia. Here the opal is found with a variety of other minerals in volcanic rocks near hot springs. Querétaro and Zimapán, Mexico, have deposits of excellent fire opal, and some precious opals have been found in Humboldt County, Nevada, and in Latah County, Idaho. Common opal occurs nearly everywhere.

QUARTZ

Quartz is one of the most widely distributed minerals in the crust of the earth. It occurs in nearly all kinds of geologic environments, and because of its durability it is one of the most common constituents of sedimentary rocks. It has a hardness of 7 and does not cleave but breaks with a glassy conchoidal fracture. The specific gravity is 2.65. Its index of refraction and dispersion power are both so low that quartz gems have little brilliance and depend upon color for their attractiveness. Quartz is such a common mineral that it is valued only as a semiprecious stone.

The varieties of quartz are almost without limit, but it is possible to recognize three different major categories: (1) Coarsely crystalline quartz often occurs in typical hexagonal crystals or in masses without external crystal form. It is transparent or transluscent and always has a glassy luster, sometimes with an oily appearance. (2) Granular fine-grained quartz has a typical dull luster like unglazed porcelain. It is exceedingly fine grained and opaque, except in very thin pieces. (3) Fibrous fine-grained quartz has a characteristic waxy luster and is commonly transluscent. Quartz in this category shows a very fine fibrous texture under the microscope, in contrast to the sugary grained appearance of the granular fine-grained quartz. The following table gives the names of the common semiprecious gems that are of each category.

GEMSTONES OF QUARTZ

Coarsely crystalline quartz (most varieties as distinct crystals):
  rock crystal—Colorless glassy quartz
  amethyst—purple or violet colored
  rose quartz—pink or rose colored, usually not as crystals
  cairngorm stone—(smoky quartz) with yellow to black smoky color
  citrine—light yellow quartz
  cat's eye—quartz with fibrous inclusions giving a silky sheen
  rutiliated quartz—with inclusions of needlelike crystals of rutile
  aventurine—quartz including sparkly flakes of mica or hematite
Granular fine-grained quartz (because of dull luster not used much as gem material):
  flint—dark in color, used by early man for various implements
  chert—same as flint only light in color
  jasper—red, yellow, or brown iron-oxide-stained chert
  prase—green-stained chert, occurs with jasper
Fibrous fine-grained quartz (*chalcedony* is a general term for this kind of quartz. Different colors and structures distinguish the following varieties. All can be artificially stained):

carnelian—red chalcedony

sard—brown chalcedony which grades into carnelian

chrysoprase—apple-green chalcedony, color due to nickel oxide

heliotrope or *bloodstone*—green chalcedony with small red spots

onyx—chalcedony with colored layers in parallel planes. A variety is *sardonyx* which has sard alternating with black or white layers.

agate—chalcedony with concentric bands in different colors and sometimes with layers of opal and other quartz varieties. *Banded agate*, with curved layers; *fortification agate*, with layers turning at sharp corners; *moss agate*, no layers, but with dark mossy growths of magnetite or manganese oxide; *agatized wood*, wood replaced or "petrified" by chalcedony

SYNTHETIC GEMS

Synthetic gems are mineral substances that are chemically and physically identical to the naturally occurring gemstones. They differ only in having been made in a laboratory. Synthetic gems can be made far more perfect than natural stones, and only an expert can distinguish them after they have been cut. Sometimes the only clues are slight differences in the growth lines, bubble holes, and small inclusions, all of which are only visible with a microscope. Rubies, sapphires, and spinels are now made as synthetic stones in great quantities. In addition to gem uses, much of the synthetic material is cut into watch and instrument jewels.

It would be well at this point to distinguish between the *synthetic gems* and the many creations that are best classified as *imitation gems*. Special glass called "paste," plastic, and common minerals are made into gems that often resemble very closely real precious stones. Of course, only a few simple tests are required to detect these imitations. Most of them are made for inexpensive costume jewerly, but from the earliest times clever imitations have been used by unscrupulous people to cheat the unwary.

The Verneuil Process

As early as 1837, small synthetic rubies had been made in the laboratory, but it was not until 1902 that a Swiss named Verneuil announced a process that could be used commercially. The apparatus (Fig. 97) is simple, being essentially an oxyhydrogen torch. The container *A* holds very pure powdered alumina which falls through a very fine sieve at the bottom of the container whenever the small mechanical hammer taps the top. Oxygen also enters

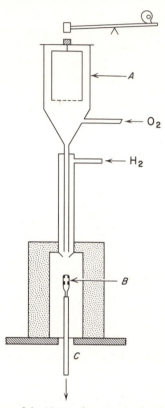

Fig. 97. Schematic diagram of the Verneuil apparatus for the growth of synthetic gem material. (From drawing provided by Linde Air Products Company.)

the apparatus through the container and the sieve. Hydrogen combines with the oxygen to give an intense flame that is directed at a clay support C. The alumina dust is melted by the intense heat and collects as a molten droplet on the clay support. As more and more molten alumina collects, the bottom of the mass cools enough to crystallize. If the clay support is gradually lowered at the same rate that alumina is collecting, a long rod- or pear-shaped mass accumulates, molten at the top and still hot but solidified at the bottom. This is called the _boule_ (B in Fig. 97). Generally, a furnace is kept in operation until a boule of 150 to 400 carats has formed (Fig. 98). This may take up to four hours. If only pure alumina is used, a colorless glassy boule of _white sapphire_ is obtained. Synthetic ruby in a deep red can be made if approximately 7% of chromium oxide is added to the alumina. A beautiful blue corundum, known as *synthetic sapphire proper,* can be made when about 2% iron oxide and 1% titanium oxide are used as coloring matter. Almost any color can be created with different combinations of pigment additives. Even a corundum gem that simulates *alexandrite* can be made with vanadium oxide as a coloring agent. Just like true alexandrite, this material is gray green in daylight and has a distinct wine-red color in artificial light. Both *star rubies* and *star sapphires* can now be made.

The Verneuil method may also be used to make *spinel* ($MgAl_2O_4$) in a variety of colors if magnesia is added in equal amounts with the alumina. Synthetic rutile (TiO_2) may also be made. It is a colorless material with an exceedingly high index of refraction and is very popular as a gemstone called *titania.*

Most of the boules from the Verneuil furnace are cooled rather rapidly and develop great internal strains. It was found that much

Fig. 98. A sapphire boule in the position where it was formed in a Verneuil furnace. Note mouth of blowpipe above and retractable clay support on which the boule has grown. (Photo courtesy of Linde Air Products Company.)

of this strain is relieved if the boule is first split lengthwise. It is then sawed into "half moon" slices that are further fabricated into jewel bearings or gems.

Most synthetic gems were made in Switzerland, France, and Germany before World War II, but now plants in England and the United States are major producers.

Synthetic Emeralds

A method to make synthetic emeralds was perfected by Carroll F. Chatham of San Francisco, California, over 20 years ago. His method is one in which the emeralds slowly crystallize from solutions. The details are still kept secret. His emeralds are of excellent quality and are difficult to distinguish from the best natural stones. Synthetic emeralds have also been produced in Germany.

Synthetic Diamonds

As long ago as 1893 Moissan created what he thought were tiny diamonds by saturating molten iron with carbon and cooling the mass rapidly. The enormous pressures created inside the ingot were thought to have changed some of the graphite into diamond. Many other interesting attempts were made, all with questionable results. Early in 1955 the General Electric Company announced that they had successfully made tiny synthetic diamonds in a huge hydraulic press where carbon was held for a few minutes at a pressure of 1.5 million pounds and at the same time heated to 5000°F.

SELECTED REFERENCES

Ball, S. H.   Geologic and geographic occurrence of precious stones, _Econ. Geol.,_ Vol. 17, pp. 575–601, 1922; also Vol. 30, pp. 630–642, 1935.

Ball, S. H. _A Roman Book on Precious Stones,_ Gemological Inst. of Am., Los Angeles, 338 pp., 1951.

Clabaugh, S. E.   Corundum deposits of Montana, _U. S. Geol. Surv. Bull._ 983, 100 pp., 1952.

DuToit, Alex. L.   _The diamondiferous gravels of Lichtenburg,_ South Africa Geol. Surv. Mem. No. 44. 50 pp., 1951.

Kraus, E. H., and C. B. Slawson. _Gems and Gem Materials,_ McGraw-Hill Book Co., New York, 1947.

Kunz, G. F.   _Gems and Precious Stones of North America,_ Scientific Publishing Co. New York, 1890.

Pearl, R. M.   _Popular Gemology,_ John Wiley and Sons, New York, 1948.

Shipley, R. M. (and others).   _Dictionary of gems and gemology, including ornamental, decorative, and curio stones,_ 5th ed., 261 pp., Gemological Inst. of Am. Los Angeles, 1951.

Williams, A. F.   _The Genesis of the Diamond,_ 2 volumes, Ernest Benn, London, 1932.

chapter 17

miscellaneous nonmetallic minerals

Mica

There are a number of minerals that belong to the group we call "mica." All of them are complex silicate minerals characterized by remarkably perfect cleavage in one direction by which crystals can be split into thin flexible transparent sheets. *Muscovite* (Fig. 99) is the so-called white mica and is the most valuable. It is a potassium-rich mineral that is colorless and transparent in thin sheets. This mica was once widely used in Russia for window panes in houses and in the portholes of war ships. It was then commonly known as "Muscovy glass," from which the technical mineral term was derived. *Isinglass* is still a popular name for this mineral. *Biotite* is "black mica," an aluminum silicate which contains magnesium and iron in addition to potassium. It is of little value but is very common in many rocks. *Phlogopite,* an amber-colored mica rich in magnesium, is valuable because it will withstand temperatures up to 1000°C. *Lepidolite* and *zinnwaldite*

Fig. 99. Muscovite mica "book" and sheets.

micas are minor sources for lithium, and *roscoelite* is a mica from which vanadium is extracted.

The most widely used mica and the one most familiar is muscovite. Unless otherwise designated, all references to commercial mica are assumed to mean muscovite.

Uses for mica

Mica is a poor conductor of electricity and hence an excellent insulator. For this reason, it is used in all vacuum tubes, coils, and condensers. Because of its ability to withstand high temperatures, mica is used to support filaments and serves as an insulator in heating appliances such as toasters, electric irons, coffee pots, waffle irons, and grills. Mica is useful for windows in ovens, stoves, and lanterns, where a substance is required that is both transparent and resistant to heat. Other uses of mica depend upon its ability to withstand chemical attacks of strong acids and caustic solutions.

Suitable electrical insulation material can be made by cementing thin splittings of mica with some bonding material such as shellac. This built-up mica plate has many uses, for it can be milled to nearly any shape. Because it wears at about the same rate as copper, it is an excellent insulator between the copper segments in motors and generators. Mica bonded to cloth tape combines high insulation properties with great flexibility.

Ground mica is made by both a wet and dry method from scrap mica that is too small for any other use. The wet-ground mica is of high quality and is used in paints, wallpaper, and inks for special sparkling effects. It is also added as a filler to rubber and some plastics. The dry-ground mica is mostly used for roofing papers, stucco, lubricants, and molded electric insulators.

Production of Mica

India has exported high-quality muscovite since 1885 from numerous deposits in the provinces of Bihar and Madras. Indian producers are also fortunate in having abundant, skilled, low-cost labor, a very important advantage in an industry so dependent upon hand operations. There are excellent deposits of mica in both Brazil and Argentina, but the product from these countries is inferior to the Indian mica because of the less skillful laborers. Mining of mica in the United States started about 1868 when a rather poor-quality sheet mica was produced. In the middle 1890's, processes were developed for grinding mica, and the American production expanded greatly to supply grinding mills. This industry prospered because it does not require the expensive hand treatment needed for sheet mica.

Geologic Occurrences of Mica

Mica occurs in many kinds of rock, but pegmatites are the main source of muscovite and yield most of the mica of economic value. It occurs as coarse crystalline masses in streaks or pockets and sometimes is found in great crystals called "books." One unusual book was found at the Purdy mine in Ontario in which the sheets were 7 x 9½ feet and the book developed to a thickness of 3 feet. It yielded about 7 tons of trimmed sheets. Scrap mica suitable for grinding can be recovered from some metamorphic rocks such as mica schist. Pegmatites in North Carolina, the Black Hills of South Dakota, and in New England yield most of the United States' production of muscovite as well as considerable lepidolite.

Phlogopite mica is a high-temperature mineral that forms when pegmatite solutions are contaminated by magnesium-rich country rock such as dolomitic limestone or pyroxenite. Deposits of this nature in Canada and Madagascar supply the world with this mineral. Some metamorphic limestones also contain phlogopite, but nowhere is there an economic occurrence of this type.

Asbestos

The Romans were the first people to use asbestos. They thought it was a vegetable fiber and named it *amianthus*. They wove it into fireproof cloth that was used to wrap the dead and hold their ashes after cremation. It was also made into lamp wicks. Marco Polo reports the use of amianthus in Siberia in the thirteenth century. No other use or mention was made of this interesting mineral until it was "rediscovered" in Italy in 1868. The discovery of the great deposit near Quebec, Canada, in 1878 started the development of asbestos as the important industrial mineral that it is today.

Asbestos is a name applied to the fibrous forms of many minerals. Of these, *chrysotile,* a variety of serpentine, is the most abundant and widely used (Fig. 100). The minerals *anthophyllite, amosite, crocidolite, tremolite,* and *actinolite* all may be fibrous enough to be considered asbestos. They belong to that group of silicate minerals known as the *amphiboles.* All of the amphibole asbestos minerals are resistant to acids and alkalis and so are useful for chemical filters. However, for the most part, their fibers

Fig. 100. Cross-fiber vein of asbestos in serpentine rock.

are brittle and lack enough strength to be spun into thread and woven into fabric. Chrysotile has a fine silky fiber that is strong and flexible and easily twisted into thread. It is unaffected by temperatures up to 450°C, and like the amphiboles it is a good electrical insulator. It, however, is easily attacked by acids.

Uses for Asbestos

Asbestos suitable for spinning is made into cloth that has a variety of uses because of its fireproof nature. Special suits, hoods, and gloves are made of asbestos cloth. These are worn by fire fighters, foundrymen, and others who may be exposed to high temperatures. Theater curtains, brake linings, gaskets, and special paper are all made of asbestos to withstand high temperatures. The shorter fiber asbestos and that not suitable for spinning may be used for insulation around electric wires. The fluffy porous fireproof fibers also make excellent heat insulation for homes. Much short fiber is mixed with cement to make asbestos siding, shingles, soil pipes, and other products widely used by builders. Asbestos is now mixed with asphalt and used in certain kinds of floor tile.

Geological Occurrence of Asbestos

The great Canadian deposits lie in a belt that extends from Vermont northeastward nearly 100 miles into southern Quebec (Fig. 101). The two main producing areas together make an area 60 to 70 miles long, and 5 to 6 miles in width. Asbestos occurs in veins which crisscross in all directions through masses of serpentine rock that are thought to be Cambrian in age. The serpentine has formed from the alteration of sills of *peridotite,* an ultrabasic rock rich in olivine. Most of the chrysotile veins are ¼ to ½ inch wide, but a few are as much as 5 inches. *Cross-fiber* veins are most common in which the individual fibers are all parallel and extend at about right angles to the walls of the vein. There is often a seam through the middle of such veins. *Slip-fiber* veins occur in fault planes and have fibers that parallel the walls of the vein. Slip-fiber chrysotile is usually rather coarse and is only good for "mill" fiber. The commercial rock yields about 5% fiber, but only about 1½% of the total fiber can be considered as spinning quality. Most of the mining started in open pits, but now the largest mines have extensive and modern underground operations. Only concrete and steel supports are used in these mines as wood

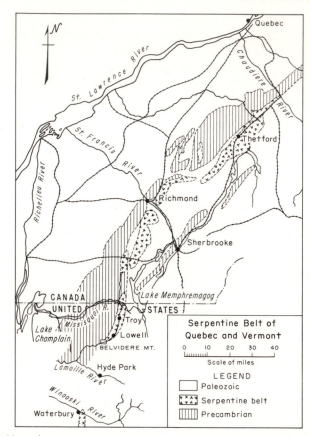

Fig. 101. Map showing serpentine areas in which are the main North American deposits of asbestos. (From Keith and Bain, *Econ. Geol.*, Vol. 27, 1932.)

timbering might contaminate the fiber with splinters that are almost impossible to remove.

Two new Canadian deposits give promise of becoming large producers within a few years. One of these is in the Yukon, about 30 miles from Dawson City, and the other near Matheson, Ontario, east of Timmins.

Some small deposits of asbestos occur in Vermont in an extension of the Quebec district. A small, but interesting deposit of a different kind was found in the Sierra Anchas in central Arizona. Here basic sills have invaded a dolomitic limestone causing some beds to be completely replaced by serpentine. In the serpentine are discontinuous seams or lenses of chrysotile from which fibers up to 6 inches long have been recovered.

The excellent deposits in *Southern Rhodesia* are similar to those in Canada. The Shabani deposit in one of the three asbestos districts occupies the central part of a serpentine mass over 10 miles long and up to three miles wide. The rock containing fiber crops out in an area 3 miles long and 600 feet wide. Much fine chrysotile is produced, of which up to 30% is spinning grade.

The *origin* of chrysotile is a subject of much controversy among those geologists who have studied asbestos. The main question is how and why does chrysotile, which has the same chemical composition as serpentine, form the characteristic veins in serpentine. Some geologists believe that the chrysotile was deposited by hydrothermal solutions as simple fillings in cracks that resulted from stresses or from hydration. Another group believes the cracks were forced open by the growth of the asbestos fibers. Still a third school of thought holds that alteration of the serpentine to chrysotile took place along the walls of solution-bearing cracks. Proponents of all these theories believe that olivine-rich igneous rocks and dolomitic limestones first alter to serpentine before asbestos is formed.

Barite

Barite, also called "heavy spar," "barytes," or "baryta" (Fig. 102), is a nonmetallic mineral with the high specific gravity of 4.5. The color is commonly white, but tints of blue, yellow, brown, gray, or red are not unusual. Barite has excellent cleavage, and its hardness varies between 2.5 and 3.5. Coarse-grained barite is commonly pearly or glassy, but the mineral sometimes occurs in very fine-grained masses that have an earthy luster. Chemically, it is barium sulfate, an insoluble and very inert substance. *Witherite,* a barium carbonate, has many of the same properties and uses as barite.

Uses for Barite

Barite is used by industry in three different forms: (1) ground barite, (2) lithopone, and (3) barium chemicals. Ground barite is employed principally by the petroleum industry for oil-well drilling mud. Where there is excessive fluid pressures in the deep strata, a column of water or clay mud may be ejected from a well,

Fig. 102.   Cluster of large barite crystals.

but muds with the high-specific-gravity barite form a column heavy enough to resist these forces.   It is common practice in such wells for the driller to use about 5 tons of ground barite for each 1000 feet of hole.   Much ground barite is used as an extender and pigment for paint and as a filler in rubber.   Much is used in the manufacture of glass where it acts as a flux and decolorizing agent. In the manufacture of paper, linoleum, oil cloth, and cardboard, barite is important as a filler and surfacing material that imparts a highly glossy surface.

Lithopone is a white pigment that consists of about 70% of barium sulfate and 30% zinc sulfide.   It is marketed as a very fine-grained powder that is chemically pure.   The crude barite is subjected to a complex process whereby it is reduced with coal in a furnace, dissolved, and reprecipitated as barium sulfate.   Zinc solutions are added that precipitate pure zinc sulfide at the same time. Lithopone is the major white pigment in paints.   It too is used as a filler in rubber and also in linoleums and textiles.   Some even is added as a component in face powder.

There are a variety of barium chemicals that are used in such diversified processes as the case-hardening of steel, the dyeing of leather and textiles, the manufacture of green signal flares, and as an indicator in X-ray photography.

Production of Barite

Barite is an abundant mineral that is marketed as a bulk material selling for a relatively low price. Consequently, there are many excellent deposits that are not worked because no large markets are available nearby. The increased use of barite drilling mud has caused the major production to be from deposits nearest oil-producing areas. The United States is the leading country in production of barite, and the most important deposits lie in Arkansas. They are centrally located with regard to the major midcontinent oil-producing areas.

Occurrence of Barite

Barite can be found in three different geologic occurrences. (1) It is a common mineral in veins and other hydrothermal fillings that have formed at relatively low temperature. In such deposits, the barite may be associated with fluorite and metallic sulfides and is considered a gangue mineral. (2) Barite occurs as isolated nodules or even as complete beds where it has replaced limestone, dolomite, shale, and other sedimentary rocks. (3) Residual deposits of barite are the result of weathering of primary deposits. Because of its insolubility, barite remains in a residual concentration when a vein, limestone, or dolomite in which it occurs gradually dissolves away. All commercial deposits of barite are either replacement or residual.

The *Arkansas* deposit at Magnet Cove near the town of Malvern is the most productive one in the United States and is an excellent example of a replacement occurrence. The barite is in a shaly limestone near the base of the Stanley shale formation of Pennsylvanian age. These rocks have been folded into a westward plunging syncline that has been cut off to the west by the intrusive rocks of the Magnet Cove, the probable source for the replacing solutions. The commercial bed varies between 40 and 75 feet in thickness and averages about 67% barium sulfate. Impurities of quartz, iron oxides, and shale are so intimately mixed that the material must be milled to a very fine state and the barite separated by flotation. The product has a specific gravity of 4.35 and is used solely for drilling mud. Reserves were estimated at over 8,000,000 tons in 1949.

In Washington County, Missouri, are extensive residual deposits of barite. At one time, before the Magnet Cove deposits were exploited, they were responsible for half of the United States' production. The primary barite occurs as veins and disseminated masses in the Potosi and Eminence dolomites of Cambrian age. Where these beds have been weathered, a red clayey soil has formed that contains about 10% barite and impurities of chert, chalcedony, limonite, and even some galena. The residual soil varies between 0 and 30 feet in thickness, and several rich concentrations or "leads" of barite are 10 to 20 feet wide and extend for several hundred feet.

SELECTED REFERENCES

MICA

Dunn, J. A. *Mica,* India Geol. Surv. Rec., Vol. 76, No. 10, 53 pp., 1947.

Harton, H. W. *Mica,* U. S. Bur. Mines. Inf. Circ. 6822, 1935.

Pecora, W. T. (and others). Mica deposits in Minas Gerais, Brazil, *U. S. Geol. Surv. Bull.* 964–c, pp. 205–305, 1950.

Sterrett, D. B. Mica deposits of the United States, *U. S. Geol. Surv. Bull.* 740, 1923.

ASBESTOS

Badollet, M. S. Asbestos, A Mineral of Unparalleled Properties, *Can. Min. and Met. Bull.,* Vol. 44, No. 468, pp. 237–246, 1951.

Bowles, Oliver. Asbestos, *U. S. Bur. Mines Bull.* 403, 92 pp., 1937.

Cooke, H. C. Asbestos deposits of Thetford District, Quebec, *Econ. Geol.,* Vol. 31, pp. 355–376, 1936.

Riordon, P. H. The genesis of asbestos in ultra basic rocks, *Econ. Geol.,* Vol. 50, No. 1, pp. 67–81, 1955.

Sinclair, W. E. The production of asbestos in South Africa, *Inst. Min. and Met. Bull.* 566, pp. 159–178, 1954.

BARITE

Dean, B. G., and D. A. Brobst. Annotated bibliography and index map of barite deposits in United States, *U. S. Geol. Surv. Bull.* 1019–C, 1955.

Harding, A. C. *Barite production in the United States,* A.I.M.E. Tech. Pub. 2412, 1948.

glossary

AERATION Exposing to the action of air. Charging with air. A process to remove objectionable gases from water supplies and sewage.

AMYGDULE A mineral-filled gas cavity in an eruptive igneous rock. It is from the Greek, and means "almond shaped."

ANORTHOSITE A coarse-grained igneous rock which consists almost entirely of plagioclase feldspar.

AQUIFER A porous and permeable rock stratum that carries water.

ARGENTIFEROUS Containing silver.

ASHLAR Facing stone for building, with rough hewn surface. It is not a dimension stone but must be cut to fit by the mason.

ASPHALTIC-BASE PETROLEUM Crude oil which leaves behind a residue of asphaltic tar after distillation.

ATOM The smallest particle of a chemical element.

BASIC A term used for igneous rocks comparatively low in silica (generally less than 50%). Usually dark-colored igneous rocks.

BATHOLITH A very large intrusive mass of coarse-grained igneous rock with no apparent bottom.

BONANZA Literally, "fair weather." A miner's term for good luck, or a body of rich ore. A mine in *bonanza* is profitably producing ore.

BRECCIA   A rock consisting of consolidated angular rock fragments larger than sand grains. It is like conglomerate, except that most of the fragments have sharp edges and unworn corners.

CALCINE   To expose a mineral substance to heat so as to drive off water or other gaseous constituents.

CAPILLARY FRINGE   Water that has been drawn upwards from the ground-water table in tiny openings and held by the surface-tension force called "capillary attraction."

CAP ROCK   A bed of relatively impervious rock, such as shale, that forms the seal above an accumulation of oil or gas. A plate of limestone, anhydrite or gypsum that lies at the top of a salt dome.

CARAT   1. Unit of weight for precious stones, standardized in 1913 to equal 0.200 gram. 2. A term used for the fineness of a gold alloy meaning ¼₄; i.e., pure gold is 24-carat; 50% gold alloy is 12-carat.

CARBOHYDRATE   A group of organic compounds of carbon, hydrogen, and oxygen. Carbohydrates are the main constituents of all plants.

CARBONATE   A chemical compound in which one or more metals are combined with the carbonate radical group consisting of one carbon and three oxygens.

CATALYSTS   Chemical substances which promote or speed up reactions between other substances but are not themselves changed or used up in the process.

COLLOIDAL PARTICLE   An exceedingly small particle with a surface electrical charge, and commonly a coating of water molecules.

CONE OF DEPRESSION   The funnel-shaped depression in the ground-water table caused by the de-watering of the soil in the vicinity of a pumping well. Its farthest extent from the well is called the _radius of influence._ Its depth at the well is called the _drawdown._

CONNATE WATER   Water deposited simultaneously with and included in the pores of sedimentary rocks and which has not since its deposition existed as surface water or atmospheric moisture.

CONTACT METAMORPHISM   A general term applied to the changes which take place along a contact (of an intruded igneous rock and the enclosing rocks into which it has been thrust) such as the recrystallization of limestone or the formation of the typical silicate minerals.

COUNTRY ROCK   The general mass of previously existing rock into which has been injected a vein or igneous intrusion.

CRACKING   A process in which the complex hydrocarbons composing petroleum or other similar oils are broken up by heat and, usually, pressure into lighter hydrocarbons of simpler molecular structure. Cracking is used in producing commercial gasoline and in enriching illuminating gas.

DRAWDOWN   The temporary downward displacement of water level in a well below the water table, caused by pumping.

DUNITE   A coarse-grained igneous rock which consists almost entirely of the mineral olivine. It is considered a variety of peridotite.

ELECTROLYTE   A substance which, when dissolved or melted, decomposes to ions in a molten fluid or a solution that will transmit an electrical current.

ELECTRON One of the smallest particles that constitutes an atom. It carries a negative charge and moves rapidly in orbit layers around the nucleus.

EPITHERMAL Pertaining to mineral deposits formed at shallow depth by deposition from low-temperature hydrothermal solutions. Certain characteristic minerals distinguish these deposits.

FIXED CARBON Chemically uncombined carbon which remains behind when coal is heated in a closed vessel until the volatile matter is driven off. It is the nonvolatile matter minus ash.

FLUX A salt or other mineral added in smelting of ores and in ceramic processes to lower the melting temperature. In metallurgical furnaces it forms a slag which contains the impurities.

FOLIATION The banding or lamination of metamorphic rocks as distinguished from the stratification of sediments, i.e., slaty cleavage and banding in gneiss.

FRACTIONAL DISTILLATION An operation for separating a mixture of two or more liquids which have different boiling points. Used extensively in petroleum distillation.

GABBRO A family of coarse-grained igneous rocks consisting of plagioclase feldspar with augite or other pyroxene minerals.

GANGUE The minerals in an ore which are of no value and are discarded.

GOSSAN The iron oxide-rich outcropping of an ore deposit caused by the weathering of pyrite and other iron-bearing minerals in the ore.

GOUGE A layer of soft pulverized rock that occurs in a fault plane. So named by miners who could "gouge" the claylike substance with a pick.

GRADE 1. The slope of the bed of a stream upon which the flow can just transport its load without either eroding or depositing. 2. A term that indicates the value of an ore with respect to its content of metal, i.e., low grade or high grade. Used also in reference to impurities in coals.

GREISEN A coarse-grained rock, composed of quartz and muscovite or some related mica, rich in fluorine. It is characteristic of the primary ores of tin. It is thought to be the result of contact action on granite by its own evolved mineralizers.

GROUND WATER TABLE The upper surface of the zone of saturation.

HEAD Pressure exerted by a body or column of fluid.

HYDRAULIC MINING A method of mining in which a bank of ore-bearing earth or gravel is washed away by a powerful jet of water and carried into sluices where the heavy ore minerals are separated from the earth by their specific gravity.

HYDROCARBONS Compounds containing only hydrogen and carbon. The simplest hydrocarbons are gases at ordinary temperatures; with increase in molecular weight they change to the liquid and finally to the solid state.

HYDROTHERMAL Pertaining to hot water and to its action in dissolving, depositing, and otherwise producing mineral changes within the crust of the earth.

HYDROXIDE A chemical compound of a metallic element or radical group combined with oxygen and hydrogen present in equal proportions and not as water.

HYPOGENE   A term proposed by Lyell for all rocks formed within the earth, now used mostly for ore deposits which were deposited by solutions that came from below.

HYPOTHERMAL   Mineral deposits formed from high-temperature hydrothermal solutions in and along deep-seated rock openings.

IONS   The positive- or negative-charged particles that result when atoms lose or gain electrons.   When in a solution or melt they can be attracted to oppositely charged electrodes.   The positive ions attracted to the cathode are called cations, and the negative ions are anions because they collect at the anode.

ISOTOPES   Atoms of the same element with essentially the same chemical properties but differing in atomic weights because of different numbers of neutrons.

JUVENILE WATER   Magmatic water from the interior of the earth which is new and has never been a part of the atmosphere.

KARST TOPOGRAPHY   Marked by sink holes, abrupt ridges, steep-walled narrow valleys, and by caverns and underground streams, in a region underlain by limestone.

LATERAL SECRETION   A theory which holds that metals in lode deposits are dissolved from the surrounding rocks by horizontally moving solutions.

LATERITE   The red residual soils or surface materials that have originated in place from the intense chemical weathering of rocks.   They are essentially characteristic of the tropics.

LATERITIZATION   Process of complete chemical weathering, whereby silicate minerals are intensely leached even of their silica, leaving concentrations of aluminum and iron oxides and hydroxides.

LIGNITE   A brownish-black coal with a woody texture.   It burns easily with a long smoky flame but has a lower heating power than other coals.

LOPOLITH   A large lenticular intrusive body of igneous rock like a sill but with a central depression.   Its upper surface is basinlike.   Lopoliths may be large bodies of several hundred square miles area, and as much as 10 miles thick.

MAGMA   A hot liquid in the earth that consists mostly of molten silicates but may contain some sulfides, phosphates, etc.   Important constituents in most magmas are great quantities of gas held in solution by pressure.

MAGMATIC DIFFERENTIATION   The process by which different types of igneous rocks, and even ore deposits, may result from the same magma, or by which different parts of a single molten mass assume different compositions and textures as it solidifies.

MAGMATIC ORE DEPOSIT   A deposit formed as part of an igneous rock.   The ore minerals crystallized from the liquid part of the magma, often as a result of magmatic differentiation.

MESOTHERMAL   Mineral deposits formed from hydrothermal solutions intermediate in temperature between epithermal and hypothermal solutions.

METALLOGENIC EPOCH   A period of time in the geologic past characterized by the extensive formation of certain metallic deposits.

METALLOGENIC PROVINCE   An extensive area characterized by the occurrence of a particular metal in a number of deposits.

METAMORPHISM   The processes whereby a rock is changed in texture and composition by abnormally intense heat and stresses.   Hydrothermal solutions may or may not play a role.

METEORIC WATER   Water that exists as atmospheric moisture or surface water; water that has entered from the surface into the voids of the lithosphere.

MINERAL DEPOSIT   Any occurrence of minerals in, or on the earth.   The term is most used where valuable minerals are present.

MIXED-BASE PETROLEUM   Crude oil which leaves a residue of both paraffin and asphalt when refined.

NEUTRON   That fundamental particle in the atomic nucleus which carries no charge and has the same mass as a proton.

NORITE   A type of gabbro containing plagioclase and orthorhombic pyroxene, usually hypersthene.

OÖLITES   Round concentric or radiating concretionary bodies in a sedimentary rock.   They are less than 2.0 mm in diameter, averaging about 1.0 mm, and may consist of many minerals.   The name means "egg stone" in Greek.

ORE   A mineral deposit from which one or more metals can be recovered profitably.   Some deposits with gems and other nonmetallic minerals are also called ores.

ORE MINERALS   The valuable constituents of an ore.   Minerals in any deposit from which valuable metals can be extracted.

OVERBURDEN   Worthless surface material covering a useful mineral deposit.

OXIDATION   The chemical process of combining with oxygen.   In a more general sense, the removal of one or more electrons from an atom or ion.

OXIDE   A compound of oxygen with one or more positive elements or radicals.

PEAT   Brown to black unconsolidated pulpy accumulations of vegetable material formed under swampy conditions.

PEGMATITE   An igneous rock, generally exceedingly coarse grained but usually irregular in texture and composition, composed mainly of quartz, feldspar, and muscovite and commonly a source of many rare minerals and gems.   Pegmatite bodies occur as dikes, lenses, or irregular bodies in or near large granitic intrusions.

PERCHED WATER TABLE   The upper surface of a lens of water which is separated from the main zone of saturation by an interval of unsaturated ground.   Such a lens owes its existence to a limited layer of impervious material.

PERIDOTITE   A granular igneous rock composed essentially of olivine, generally with some form of pyroxene and with or without hornblende, biotite, chromite, garnet, etc.

PERMEABILITY   It is the measure of the ease or difficulty with which a rock will allow the movement of a fluid through itself.

PHONOLITE   A fine-grained igneous rock consisting of essential orthoclase and nepheline, with accessory amphibole, pyroxene, or mica.   Nepheline syenite has the same minerals but is coarser.

PHOSPHATE   Chemical compound of one or more positive elements with the phosphate radical of one phosphorus and four oxygen atoms.

PIEZOMETRIC SURFACE   The piezometric surface of an aquifer is an imaginary plane that everywhere coincides with the level to which the water from a given aquifer will rise under its full head.

PISOLITES   Globular concretions about the size of a pea.   Concretionary masses that occur commonly in some limestones, clay, and bauxite. They are larger than 2 mm in diameter (the upper limit for oölites).

PLAYA   A basin of deposition in a valley with no drainage outlet, found only in arid regions.   Playas often contain salt lakes or salt-encrusted flats.

POLYMERIZATION   Process whereby light organic compounds can be joined to make larger more complex molecules.   The opposite of cracking in petroleum refining.

POROSITY   The volume of pore space expressed as a percentage of the total volume of the rock mass.

PORPHYRY   An igneous rock which has a scattering of completely formed crystals in a fine-grained matrix.   The crystals are called _phenocrysts_.

"PORPHYRY COPPER"   Name applied to the many copper deposits like many in the western states in which the ore minerals are disseminated throughout a large mass of highly fractured igneous rock.

PRIMARY   Refers to unweathered minerals and ore; thus, the first or original minerals in a deposit.

PROTONS   A fundamental particle in the atomic nucleus which carries a positive charge as strong as the negative charge of the electron.

PROTORE   Low-grade mineral deposits which through weathering or other natural surface processes can be so concentrated to become ore.

PROXIMATE ANALYSIS   The determination of the compounds contained in a mixture as distinguished from ultimate analysis, which is the determination of the _elements_ contained in a compound.   Used in the analysis of coal. (Fay)

PYROCLASTIC   Fragmental volcanic material that has been ejected in explosive eruptions.   Rocks composed of such material are also called pyroclastics.

PYROXENITE   A coarse-grained igneous rock, consisting essentially of pyroxene, with or without hornblende, spinel, and iron oxides, and with little or no feldspar or olivine.

RADIUS OF INFLUENCE   The distance from a pumping well within which any depression of the ground-water table occurs.

RECRYSTALLIZATION   A process, generally in metamorphism, whereby fine-grained rocks are altered to become coarse-grained rocks of the same mineral content, i.e., limestone changes to coarse-grained marble.

REDUCING AGENT   A substance in a chemical reaction that causes metallic ions to lose their positive charge and become uncombined metal atoms; i.e., coke (carbon) is a reducing agent changing iron oxide to molten metallic iron in a blast furnace.

REDUCTION   A chemical process of removing oxygen.   The procedure of recovering metals from their ore minerals, either from solutions or from melts.   The gain of electrons by ions or atoms.

REFRACTORY   Having the ability to resist heat, melting only at very high temperatures.

REPLACEMENT   A process by which a mineral takes the place of an earlier different substance, often preserving the structure or crystalline form of the original material. (Fay)

RHYOLITE   A fine-grained light-colored igneous rock made up of orthoclase, quartz, and usually biotite. It commonly forms in near-surface intrusions and lava flows.

SADDLE REEFS   Ore deposits which made saddle-shaped lenses at the crests of anticlinal folds; *inverted saddles* are similar features in synclines.

SECONDARY   Refers to minerals or deposits that have formed by alteration of the primary ore, usually by weathering.

SERPENTINE   A metamorphic rock composed chiefly or wholly of the mineral serpentine. It forms from a rock rich in magnesium.

SILICATES   Chemical compounds in which silicon and oxygen in various proportions make the negative radical group.

SILL   An intrusive sheet of igneous rock of approximately uniform thickness, which has been forced between level or gently inclined beds. (Fay)

SKARN   Contact metamorphic rock of coarse carbonates and silicates of calcium and magnesium with introduced minerals of silica, aluminum, iron, and other metals.

STOCKWORK   A mineral deposit which occurs in the closely spaced interlacing cracks of a highly shattered mass of rock. The whole fractured zone may be mined as a large low-grade ore deposit.

STYLOLITE   Irregular wavy joints and bedding planes in limestones, marked by dark clay and caused by solution. They rarely occur in other kinds of rock.

SULFATE   Chemical compound in which one sulfur and four oxygen atoms make up the negative radical group.

SULFIDE   Chemical compound of one or more metals or positive radical combined with sulfur.

SUPERGENE   A term applied to minerals deposited from downward-moving surface waters, as opposed to hypogene.

SUPERGENE ENRICHMENT   A process whereby metals dissolved by downward-moving water from a mineral deposit above the water table are redeposited below the water table, enriching the ore in this zone.

SYENITE   A coarse-grained igneous rock composed essentially of orthoclase and usually biotite or hornblende, also with microcline, albite, augite, or corundum.

TACONITE   Mostly an iron-rich banded chert, but with some carbonate layers. Taconite forms the protore for the iron deposits on the Mesabi Range, Minnesota. Taconite rich in magnetite can now be mined as ore.

TECTONIC BRECCIA   A rock composed of coarse, angular fragments that have been broken by folding or faulting.

TELLURIDE   A chemical compound of a metal with tellurium. Some rare gold and silver minerals are tellurides.

TENOR   The percentage or average metallic content of an ore, matte, or impure metal. (Fay)

ULTRABASIC   A term applied to those dark igneous rocks that crystallized from magmas with very little silica.   Olivine is a common constituent, as well as many sulfide and oxide ore minerals.   Feldspars are mostly absent.

VOLATILIZE   To change to a vapor usually as a result of heating.

VUG   A cavity, lined with crystals, that is present in a sedimentary rock, in a vein, or other hydrothermal deposit.

WEATHERING   Process of physical and chemical disintegration of rocks and minerals when subjected to the agents of the atmosphere.

REFERENCES

The following extensive glossaries were consulted in the preparation of the above list and are excellent sources for the meaning of any technical geologic term.

Fay, Albert H.   _A Glossary of the Mining and Mineral Industry,_ U. S. Bur. Mines Bull. 95, 2nd printing, 1948.

Rice, C. M.   _Dictionary of Geological Terms,_ Princeton, New Jersey, 1947.

Stokes, W. L., and D. J. Varnes.   _Glossary of Selected Geologic Terms,_ Colorado Scientific Society Proceedings, Vol. 16, 1955.

Howell, J. V.   _Glossary of Geology and Related Sciences,_ The American Geological Institute, Washington, D. C., 1957.

appendix II list of the
chemical elements

appendix II

LIST OF THE CHEMICAL ELEMENTS

A	Argon (I)*	Ge	Germanium (M)	Pt	Platinum (M)
Ac	Actinium (M)	H	Hydrogen (N)	Pu	Plutonium (M)
Ag	Silver (M)	He	Helium (I)	Ra	Radium (M)
Al	Aluminum (M)	Hf	Hafnium (M)	Rb	Rubidium (M)
Am	Americium (M)	Hg	Mercury (M)	Re	Rhenium (M)
As	Arsenic (N)	Ho	Holmium (M)	Rh	Rhodium (M)
At	Astatine (N)	I	Iodine (N)	Rn	Radon (I)
Au	Gold (M)	In	Indium (M)	Ru	Ruthenium (M)
B	Boron (N)	Ir	Iridium (M)	S	Sulfur (N)
Ba	Barium (M)	K	Potassium (M)	Sb	Antimony (M)
Be	Beryllium (M)	Kr	Krypton (I)	Sc	Scandium (M)
Bi	Bismuth (M)	La	Lanthanum (M)	Se	Selenium (N)
Bk	Berkelium (M)	Li	Lithium (M)	Si	Silicon (N)
Br	Bromine (N)	Lu	Lutetium (M)	Sm	Samarium (M)
C	Carbon (N)	Mg	Magnesium (M)	Sn	Tin (M)
Ca	Calcium (M)	Mn	Manganese (M)	Sr	Strontium (M)
Cd	Cadmium (M)	Mo	Molybdenum (M)	Ta	Tantalum (M)
Ce	Cerium (M)	N	Nitrogen (N)	Tb	Terbium (M)
Cf	Californium (M)	Na	Sodium (M)	Tc	Technetium (M)
Cl	Chlorine (N)	Nb	Niobium (M)	Te	Tellurium (N)
Cm	Curium (M)	Nd	Neodymium (M)	Th	Thorium (M)
Co	Cobalt (M)	Ne	Neon (I)	Ti	Titanium (M)
Cr	Chromium (M)	Ni	Nickel (M)	Tl	Thallium (M)
Cs	Cesium (M)	Np	Neptunium (M)	Tm	Thulium (M)
Cu	Copper (M)	O	Oxygen (N)	U	Uranium (M)
Dy	Dysprosium (M)	Os	Osmium (M)	V	Vanadium (M)
Er	Erbium (M)	P	Phosphorus (N)	W	Tungsten (M)
Eu	Europium (M)	Pa	Protactinium (M)	Xe	Xenon (I)
F	Fluorine (N)	Pb	Lead (M)	Y	Yttrium (M)
Fe	Iron (M)	Pd	Palladium (M)	Yb	Ytterbium (M)
Fr	Francium (M)	Pm	Promethium (M)	Zn	Zinc (M)
Ga	Gallium (M)	Po	Polonium (M)	Zr	Zirconium (M)
Gd	Gadolinium (M)	Pr	Praseodymium (M)		

* (I) means inert gas, (M) means metal, and (N) means nonmetal

appendix III

geologic
time
scale

GEOLOGIC TIME SCALE   Lengths of Time in Millions of Years

Era	Period	Epoch	(million years)	Duration	Percentage / Started
Cenozoic Era	Quaternary Period	Recent Epoch	1	60 million years duration	3% of geologic time. Started 60 million yrs. ago
		Pleistocene Epoch			
	Tertiary Period	Pliocene Epoch	11		
		Miocene Epoch	16		
		Oligocene Epoch	12		
		Eocene Epoch	20		
		Paleocene Epoch			
Mesozoic Era	Cretaceous Period		70	125 million years duration	7% of geologic time. Started 185 million yrs. ago
	Jurassic Period		25		
	Triassic Period		30		
Paleozoic Era	Permian Period		25	335 million years duration	17% of geologic time. Started 520 million yrs. ago
	Pennsylvanian Period		25		
	Mississippian Period		30		
	Devonian Period		55		
	Silurian Period		40		
	Ordovician Period		80		
	Cambrian Period		80		
Proterozoic Era / Archeozoic Era	Called Precambrian Time		Oldest age determined on a rock: 3300 million years	3480 (?) million years duration	73% of geologic time

Origin of the earth   ? ? ? ? ? ? ?   4000 to 5000 million years ago   ? ? ? ? ? ? ?

? ? ? ? ? ? ?

list
of the ore minerals
# of various metals

ALUMINUM
bauxite   (Al hydroxide mixture)
cryolite Na_3AlF_6

ANTIMONY
stibnite Sb_2S_3

ARSENIC
arsenopyrite   FeAsS
realgar   AsS
orpiment As_2S_3

BERYLLIUM
beryl $Be_3Al_2Si_6O_{18}$

BISMUTH
native bismuth   Bi
bismuthinite Bi_2S_3
bismite Bi_2O_3

CADMIUM
greenockite   CdS

CHROMIUM
chromite $FeCr_2O_4$

COBALT
cobaltite   CoAsS
smaltite $CoAs_3$
linnaeite Co_3S_4
erythrite $Co_3As_2O_8 \cdot 8H_2O$

COPPER
native copper   Cu
cuprite Cu_2O
chalcocite Cu_2S
covellite   CuS
bornite Cu_5FeS_4
chalcopyrite $CuFeS_2$
enargite Cu_3AsS_4
malachite $Cu_2CO_3(OH)_2$
azurite $Cu_3(CO_3)_2(OH)_2$

GOLD
native gold   Au
sylvanite $AuAgTe_4$
calaverite $AuTe_2$

323

IRON
hematite Fe_2O_3
magnetite Fe_3O_4
limonite $FeO(OH) \cdot nH_2O$
siderite $FeCO_3$

LEAD
galena   PbS
cerussite $PbCO_3$
anglesite $PbSO_4$

MAGNESIUM
magnesite $MgCO_3$
dolomite $CaMg(CO_3)_2$
brucite $Mg(OH)_2$

MANGANESE
pyrolusite MnO_2
manganite $Mn_2O_3 \cdot H_2O$
psilomelane $MnO \cdot MnO_2 \cdot 2H_2O$
hausmannite Mn_3O_4
rhodochrosite $MnCO_3$

MERCURY
cinnabar   HgS
metacinnabarite   HgS
livingstonite $HgSb_4S_7$
native mercury   Hg

MOLYBDENUM
molybdenite MoS_2
wulfenite $PbMoO_4$

NICKEL
pentlandite   (Fe,Ni)S
millerite   NiS
niccolite   NiAs
annabergite $Ni_3As_2O_8 \cdot 8H_2O$
garnierite $(Ni,Mg)SiO_3 \cdot nH_2O$

PLATINUM
native platinum   Pt
sperrylite $PtAs_2$

SILVER
native silver   Ag
cerargyrite   AgCl

argentite Ag_2S
polybasite $Ag_2S \cdot Sb_2S_3$
proustite Ag_3AsS_3
pyrargyrite Ag_3SbS_3

TANTALUM and COLUMBIUM (NIOBIUM)
columbite $(Fe,Mn)(Nb,Ta)_2O_6$
tantalite $(Fe,Mn)Ta_2O_6$

TIN
cassiterite SnO_2
stannite Cu_2FeSnS_4

TITANIUM
rutile TiO_2
ilmenite $FeTiO_3$

TUNGSTEN
scheelite $CaWO_4$
wolframite $(Fe,Mn)WO_4$

URANIUM
pitchblende   (U oxide mixture)
uraninite   U oxides
carnotite $K_2(UO_2)_2(VO_4)_2 \cdot nH_2O$
tyuyamunite $CaO \cdot 2UO_3 \cdot V_2O_5 \cdot nH_2O$
autunite $CaO \cdot UO_3 \cdot P_2O_5 \cdot nH_2O$
torbernite $CuO \cdot UO_3 \cdot P_2O_5 \cdot nH_2O$
uranophane $CaO \cdot 2UO_3 \cdot 2SiO_2 \cdot 6H_2O$
gummite $UO_3 \cdot nH_2O$ with Pb, Ca, Ba, etc.
brannerite   (U and Ti oxide with rare earths, Ca, Th, Fe)

VANADIUM
carnotite $K_2(UO_2)_2(VO_4)_2 \cdot nH_2O$
roscoelite $K_2V_4Al_2Si_6O_{20}(OH)_4$
vanadinite $Pb_5Cl(VO_4)_3$
patronite VS_4

ZINC
sphalerite   ZnS
smithsonite $ZnCO_3$
hemimorphite $Zn_4Si_2O_7(OH)_2 \cdot 2H_2O$
zincite   ZnO
franklinite $(Fe,Zn,Mn)(Fe,Mn)_2O_4$
willemite Zn_2SiO_4

list
of valuable
nonmetallic minerals

amosite $(Fe,Mg)_7Si_8O_{22}(OH)_2$, amphibole asbestos
anhydrite $CaSO_4$, common impurity with salt, sulfur, gypsum
anthophyllite $(Mg,Fe)_7Si_8O_{22}(OH)_2$, amphibole asbestos
apatite $Ca_5(F,Cl)(PO_4)_3$, phosphate source
barite $BaSO_4$, barium chemicals, drilling muds
beryl $Be_3Al_2Si_6O_{18}$, beryllium metal, emerald, aquamarine
biotite $K(Mg,Fe)_3AlSi_3O_{10}(OH)_2$, black mica
borax (tincal) $Na_2O \cdot 2B_2O_3 \cdot 10H_2O$, boron chemical, flux, cleanser
calcite $CaCO_3$, mineral of limestone and marble, flux, lime
carnallite $KCl \cdot MgCl_2 \cdot 6H_2O$, potash source
chalcedony SiO_2, one of quartz semiprecious gemstones (see page 296)
chrysotile $Mg_3Si_2O_5(OH)_4$, serpentine asbestos, most valuable kind
colemanite $2CaO \cdot 3B_2O_3 \cdot 5H_2O$, source of boron chemicals
crocidolite $Na_3Fe''_3Fe'''_2Si_8O_{22}(OH)_2$, blue amphibole asbestos
dolomite $CaMg(CO_3)_2$, common mineral in limestone, building stone
fluorite CaF_2, source of fluorine, hydrofluoric acid, flux
Gem minerals (see table on page 287)
graphite   C, electrodes, lubricants, pencil leads, etc.

gypsum $CaSO_4 \cdot 2H_2O$, plaster of Paris, alabaster, portland cement

halite NaCl, common salt, important chemical mineral

kainite $MgSO_4 \cdot KCl \cdot 3H_2O$, source of potash

kernite $Na_2O \cdot 2B_2O_3 \cdot 4H_2O$, source of boron chemicals

langbeinite $2MgSO_4 \cdot K_2SO_4$, source of potash

lepidolite $K_2Li_3Al_4Si_7O_{21}(OH,F)_3$, a mica and source of lithium

muscovite $KAl_2AlSi_3O_{10}(OH)_2$, white mica

phlogopite $KMg_3AlSi_3O_{10}(OH)_2$, brown mica

polyhalite $K_2SO_4 \cdot MgSO_4 \cdot 2CaSO_4 \cdot 2H_2O$, source of potash

priceite $4CaO \cdot 5B_2O_3 \cdot 7H_2O$, source of boron chemicals

pyrophyllite $Al_2Si_4O_{10}(OH)_2$, used for same things as talc

quartz SiO_2, piezoelectric oscillators, many gems (see page 297)

sassolite H_3BO_3, natural boric acid

sericite a fine-grained flaky variety of muscovite

serpentine $Mg_3Si_2O_5(OH)_4$, main mineral of serpentine marble

spodumene $LiAlSi_2O_6$, main source of lithium, gem (see page 287)

sylvite KCl, source of potash

talc $Mg_3Si_4O_{10}(OH)_2$, ceramic material, filler, talcum powder

tourmaline (complex silicate mineral) gemstone (see page 287)

tremolite $Ca_2Mg_5Si_8O_{22}(OH)_2$, poor amphibole asbestos

ulexite $Na_2O \cdot 2CaO \cdot 5B_2O_3 \cdot 16H_2O$, source of boron chemicals

vermiculite (altered micas) expands on heating, insulation

witherite $BaCO_3$, source of barium, used like barite

zeolite silicate family synthetic zeolite used for water softener

zinnwaldite an iron-lithium mica similar to biotite

<table>
<tr><td>appendix VI</td><td># general
references</td></tr>
</table>

Adams, F. D. *The Birth and Development of the Geological Sciences,* Dover Publications, New York, 1938.

Bateman, A. M. *Economic Mineral Deposits,* 2nd ed., John Wiley and Sons, New York, 1950.

Dolbear, S. H., and O. Bowles (editors). *Industrial Minerals and Rocks,* 2nd ed., Seeley W. Mudd Series, Am. Inst. of Min. and Met. Eng., New York, 1949.

Emmons, W. C. *Principles of Economic Geology,* 2nd ed. McGraw-Hill Book Co., New York, 1940.

Encyclopedia Britannica

Ford, W. E. *Dana's Textbook of Mineralogy,* 4th ed., John Wiley and Sons, New York, 1932.

Geology and Economic Minerals of Canada, Geol. Surv. of Canada, 1957.

Hurlbut, C. S. *Dana's Manual of Mineralogy,* 17th ed., John Wiley and Sons, New York, 1959.

Ladoo, R. B., and W. M. Myers. *Nonmetallic Minerals,* 2nd ed., McGraw-Hill Book Co., New York, 1951.

Leith, C. K. *The Economic Aspects of Geology,* Henry Holt and Co., New York, 1921.

Lilley, E. R. *Economic Geology of Mineral Deposits,* Henry Holt and Co., New York, 1936.

Lindgren, Waldemar. *Mineral Deposits,* McGraw-Hill Book Co., New York, 1933.

327

Lovering, T. S. *Minerals in World Affairs,* Prentice-Hall, Englewood Cliffs, 1943.

Minerals Yearbook, annual publication, U. S. Bureau of Mines, Washington, D. C.

Newhouse, W. H. (editor). *Ore Deposits as Related to Structural Features,* Princeton University Press, Princeton, 1942.

Ore Deposits of the Western United States—Lindgren Volume, Am. Inst. Min. Met. Eng., New York, 1933.

Ries, H. *Economic Geology,* 7th ed., John Wiley and Sons, New York, 1937.

Structural Geology of Canadian Ore Deposits, a Symposium, Can. Inst. Min. and Met., Montreal, 1948.

Van Royan, W., and O. Bowles. *The Atlas of Mineral Resources of the World,* Prentice-Hall, Englewood Cliffs, 1952.

INDEX

index